Elco PT Boats: The Complete History

Frank J. Andruss Sr.

Nimble Books LLC

ISBN: 978-1-60888-526-8

Printed in the United States of America

Foreword

When I was asked to write the foreword for this book, I was both surprised and honored to do so.

I have always enjoyed talking about the Elco Naval Division, having lived next door to the facility almost all of my life, starting in October 1943 to the present. I worked for the Electro Dynamic Company, which during the war was making generators for the Navy Department and the PT boats.

I was employed there for forty years from 1956 until 1996 when I retired. When Elco closed its doors in 1949, many of their workers came to work at the Electro Dynamic Company, a large plant that was located at 163 Avenue A, in Bayonne, New Jersey. Over the years I spoke with many of those workers about their time spent building the PT boats, and its operations, including management and workers. I also had an opportunity to see and read a lot of papers, old brochures, etc. This all helped in my knowledge about the plant, and some of the people that worked there during the war years.

Growing up in this neighborhood, I knew many of the workers, both men and women, who were very skilled and dedicated. Elco initially was called the Electric Launch Company, which later would be called Elco Motor Yachts, which built electric launches and many other pleasure boats. Prior to World War II, Elco began their interest in motor torpedo boats and would build a one-million-dollar state-of-the-art plant that would be called the Elco Naval Division.

One large part of the success of the PT boat program was Irwin Chase, a graduate of the University of Michigan. He started with Elco in 1906 as a Navy architect and worked his way up to managing constructor. Chase loved speed, and the key to the success of the PT program were the hull designs, which worked well when combined with high-speed engines.

When one thinks about the massive effort it took to build 399 PT boats during the war years, it boggles the mind. The amount of raw materials being delivered to Elco by rail and ship each day was staggering. Dedicated workers were a huge reason for the success of the company. They worked side by side building a boat that Navy men would be proud of.

This book is a great read for anyone wishing to learn more about this once-great company. The photographs contained in this book are priceless and help keep the reader's attention. In going through the many pages, I can't help but have a feeling of anticipation. There is no doubt in my mind that Elco is very much alive today, one reason why Frank's book is so important to the history of the PT boats.

—Andrew J. Shanahan Jr.

Preface

It is hard to believe that this is my fifth book that deals with the World War II PT boats. I never intended to continue producing the books, but somehow my passion just took over.

With this new book I am delighted having come across so many new and unpublished photos. Elco was the front-runner in PT boat production during the war years and did an incredible job of documenting their boats, both on film and video. In doing this book it is my intention to help educate the public about not only the role of the Elco PT boat but the workers and management that lived the war years producing them.

Running twenty-four hours a day, the workmanship in building these wooden boats was incredible. Nothing left the plant unless it was 100 percent, a true testament to the people that built them. During the war the plant at Elco would receive the Army-Navy E Award for excellence six times. It is in these photos that the book comes alive, as you will see firsthand the many stages of assembly.

I firmly believe that to tell a story, one needs to see photos and not just a mass of text. Here, you will see history firsthand in photos as well as chapters that explain everything dealing with the history of these fine boats. During the war these little midget sluggers were on the front lines in all theaters of war, taking the fight to the enemy and proving their worth, not only as torpedo boats but later as gunboats, making the Elco PT a dangerous foe.

It is my hope that you the reader will become educated as to the Elco Naval Division in Bayonne, New Jersey, and her fine history.

Contents

Foreword iii

Preface v

1 Scott-Paine Design and the Elco Seventy-Foot Boat 3

2 Elco's Seventy-Seven-Foot Design 55

3 Elco Eighty-Foot 109

4 The Workers That Made It Happen 183

5 The Lighter Side of War 243

6 End of an Era 305

Acknowledgements 335

Chapter 1

Scott-Paine Design and the Elco Seventy-Foot Boat

The Electric Launch Company was one of the world's premier builders of powerboats. In its stellar fifty-seven years of boat building, Elco produced more than six thousand pleasure boats and more than fifteen hundred military vessels, from the fifty-five electric launches introduced to the world at the 1893 Chicago World's Colombian Exposition, to the 399 PT boats of World War II. The man who managed Elco was Executive Vice President Henry R. Sutphen, a minister's son from Morristown, New Jersey, and an electrician by trade. It appears Sutphen joined Elco as early as 1892, and may have been the brains behind the World's Fair project. In 1905, the University of Michigan graduated a naval architecture student named Irwin Chase, who Sutphen would hire in 1906.

Irwin Chase was a powerboat man to the core. He particularly loved to experiment with high-powered boats, and is credited (at least by Elco) with designing one of the world's first "planing hulls", a twenty-foot "Elcoplane" named "Bug", in 1911. Chase spent countless hours on the water developing high-speed small boats, and like other designers of his time his experiments involved "stepped" hulls. He designed a twenty-footer with a perfectly flat bottom, and then corrugated metal sheets to the after half. Using special shims, Chase could alter the angles of the boat's steps, and he also experimented with step angles to correct the boat's tendency to slip sideways on high-speed turns.

Sutphen promoted Irwin Chase to the position of general manager, and named Glenville S. Tremaine chief designer.

Tremaine had been a ship fitter's apprentice at the Bath Iron Works in his native Maine, and he took correspondence courses in draftsmanship at the same time. He studied yacht design under Henry Douglas Bacon and was "loaned" to naval architect Morris Whitaker for a brief assignment to New York in 1912. When work at Bath slowed down, Tremaine answered a newspaper ad for a draftsman and was hired by Elco in March, 1913, on the day Woodrow Wilson was inaugurated. The following day, Tremaine turned twenty-one.

Tremaine was Chase's apprentice and assistant draftsman until the 1923 shake-up; after his promotion. Tremaine carried on the Elco design traditions that Chase began in 1906, and he drew wonderful cruisers and motor yachts through the Roaring Twenties, aided by his new assistant, Alfred "Bill" Fleming. Tremaine loved to create new boat designs, and he loved tinkering with new ideas; he had little patience with details, however, and left that to Fleming. In time, Fleming became Elco's premier stylist.

In 1923, Elco opened its famous "Port Elco" showroom in downtown New York, and for the first time customers could walk in off the street and browse among a selection of yachts from thirty-two feet to more than fifty feet in length. Henry Sutphen not only borrowed standardized construction from the auto industry but began to adapt Detroit marketing techniques as well. In that year, Sutphen also shook up Elco's management. Before leaving town on a business trip, he had assigned Elco General Manager Thomas Hanson the task of preparing new boats for the 1923 Motorboat Show. When Sutphen returned to Elco, he found out that little or nothing had been done for the show, so he fired Hanson. Sutphen then promoted Irwin Chase to the position of general manager, and named Glenville S. Tremaine chief designer.

The Depression hurt Elco, as it hurt most boat builders, but the effects of the Depression didn't make themselves felt right away. Elco continued to build boats and pay for eight-page advertising spreads in boating magazines well after 1929. The low point came in 1933, when Elco was reduced to only three full-time employees: Chase, Tremaine, and general foreman Charles Lamont. To keep busy, they personally fulfilled orders for spare parts, answered correspondence, and ran exhaustive dynamometer tests on new engines, just for something to tinker with. Gradually, over several years, the old employees were rehired and boat building resumed.

Franklin Delano Roosevelt's relationship with Elco dated back to the World's Colombian Exposition in 1893, when Roosevelt's father took him to the fair. Young Franklin took one of his first boat rides in one of the electric launches, and may have developed his lifelong interest in things nautical from this experience. During World War I Roosevelt was Assistant Secretary of the Navy, and had worked closely with Sutphen and Chase on the motor launch project. Sutphen and Roosevelt maintained their friendship through Roosevelt's Presidency.

It was in 1938 that Elco, through the efforts of Charles Edison, at the time Secretary of the Navy became interested in the possibilities of producing torpedo boats for the US Navy.

Sutphen kept a close eye on European development and Hurbert Scott Paine, who was recognized as a genius. At the time he was employed by the British Power Boat Company. At the time the British had the Thorneycroft from the CBM, the Vosper, and Scott Paine designed boats. Elco felt that if one of these designs could be purchased, or at least an American license for one of these designs, there would be a market for the craft with the US Government.

It was on February 10, 1939 that Sutphen and at the time Chief Designer, Irwin Chase embarked for England onboard the Queen Mary, to check out the boats. What made Elco have such a sudden interest in PTs was summed up by Chase when he said, the principle reason was we had a long experience in hull design, with which the average Naval Officer was not familiar with. We felt that a small high-speed motor boat could be so designed so that she would be fit to head to sea, and stay there in any type of weather. Chase based Elco's gamble on their superior knowledge and ability to handle the factors of rough weather and high-speed engines in relationship to the boats.

Sutphen and Chase were given unusual opportunities by the Royal Navy to observe the boats. They both felt that the US would be better off with one of the boats, but choosing one was most difficult.

Both men really wanted to pick a boat that could handle the roughest treatment. This would be the turning factor in their decision to pick the Scott-Paine seventy-foot design. It was then and there that Elco would make the choice to purchase one of the boats. Negotiations for the Scott-Paine boat were successful, and on March 17, 1939 Mr. Sutphen reported by radio that the boat had been made and that favorable terms to ship the boat in three months would be made. In order to comply with the wording of the appropriation act, Elco on April 25, 1939 submitted proposal for construction of one seventy-foot Scott-Paine design. This contract was signed on June 1, 1939.

The boat was shipped to Elco in New York on September 4, 1939. From there it would be lightered to the Electric Boat Company plant in Groton, Connecticut. Dubbed PT-9, the boat was put through a series of tests and sea trials. Commodore A. Loring Swasey acted as observer for these trials. Scott Paine himself came over with the boat, and was in charge of its operation.

Elco would invite the US Navy to send their representatives, to take trial spins on the boat. The Coast Guard would also send an observer for the trials, although they felt this type of craft was too limited to warrant an investment. Scott-Paine's performance with PT-9 was very influential in the ultimate sale of the boat to the government.

When PT-9 arrived in Bayonne, it was powered by three Rolls Royce Merlin engines. Col. J.G. Vincent, Packard's chief designed felt that an American engine could better the performance of the Rolls Royce product. He convinced Elco, and later PT-9 was re-engined using the Packard W-8 engine. The boat displaced some forty tons, carrying four eighteen-inch torpedoes in four front launching tubes. Government acquisition was slated for November 1, 1939, but this date was postponed until January 3, 1940. This worked out for Elco as they used this delay to familiarize themselves with the boat. Later, acquisition was delayed six months, which proved to be in Elco's favor, as they studied construction and were prepared to build their own version of the boat.

Delivery finally took place on June 17, 1940, and PT-9 was placed in service with Motor Torpedo Boat Squadron One on July 24, 1940. Elco based its first boats on the Scott-Paine model and the delivery delay allowed them to examine the boat, check out its construction principles, and adapt or modify the design characteristics of the English boat to best suit the needs of our Navy. Of course not all the time was spent in Bayonne, as they did underway sea testing and ordnance material tests.

On November 10, 1939, Elco was allotted five million dollars to build boata. Sue to difficulties Elco was having with Packard in contracting for twenty-three additional boats, there was a small delay which was ironed out and on December 7, 1939, the Navy made an award to build eleven Motor Torpedo Boats, PT's 10–19, and twelve Motor Boat Sub Chasers, PTCs 1–12.

Elco began construction of the seventy-foot boats in January of 1940, at their plant in Bayonne, New Jersey. The new boats were basically a modified version of PT-9. As mentioned the three engines would be powered by Packard Marine Engines, powering some twelve hundred hor each. The canopy was a bit larger and could be removed in three sections. The pilot house was enlarged and flush with the trunk canopy. This was done to try and eliminate sea spray and obtain better wind deflection.

A storage space was added to stow gear by fitting an alloy boot to the front of the pilot house below her forward windows. The bridge was enlarged and lined with aluminum on the sides and the deck. Aluminum grab rails were installed and fitted around the bridge and top of the canopy. The Dewandre-powered machine gun turrets were enlarged, located mid section of the canopy, and could now hold .50 caliber Browning M-2 machine guns, instead of the smaller .30 caliber machine guns. Six cowl vents were added to the aft section of the deck.

The boats were constructed using an Alaskan spruce keel, with the frames being made using African mahogany, with sides and bottom double planked mahogany, with canvas and marine glue, in between the diagonally spaced planking. They used monel screws below the waterline, with brass or bronze above. None of these boats ever saw combat with the US Navy, as they were lend-leased to the British. They did see considerable action with the British. Elco made major modifications to the boats before being turned over to bring them to British specifications.

It was felt that these boats were too short to carry the American twenty-one-inch torpedoes. It was because of this, that the Bureau of Ordnance recommended that all succeeding PT boats would enlarged in length, which would result in Elco's seventy-seven-foot PT boat.

Figure 1-1. *Henry R. Sutphen was born in Morristown New Jersey, a son of a Presbyterian minister. He would attend school at Morris University and New York University. He did pioneer work in the development of the electric launch and in the application of gasoline engines in marine craft. Sutphen joined Elco in 1892,where he initiated the design of the submarine chasers and in April 1915 signed a contract with the British for fifty 80 foot Motor Launches. Before the outbreak of World War Two, he brought interest of the Patrol Torpedo Boat to the United States Navy. It would be under his supervision that the PT boats were built for the Navy. Long prominent in the ship building industry, Sutphen was the Executive Vice President of the Elco Naval Division of the Electric Boat Company in Bayonne, New Jersey. He was also the director of the Sutphen Product Corporation and of Canadair LTD, of Canada. Sutphen was 75 years old when he died of a stroke in his home at 876 Park Avenue in New York. (Bert Morgan)*

Figure 1-2. *Irwin Chase was considered to be one of the "founding fathers" of the extremely successful American PT boat program. Chase, a University of Michigan Naval architecture graduate of 1905, became fascinated early in his career with small, fast power boats and according to Elco, created the first "planing" hulled power boat. In 1915, Chase is also credited for designing the Elco Cruisette, then considered to be the world's most popular cruising power boat during the years between the two world wars. In February of 1939, Elco's Henry R Sutphen asked Mr. Chase to accompany him on the Queen Mary at the request of President Roosevelt (FDR) to evaluate and purchase at their discretion one of England's motor torpedo boats. Irwin Chase was a critical part of the design team that allowed Elco to become the largest manufacturer of American PT boats at its newly built extensive facility at Bayonne, New Jersey. Eventually he transferred from Bayonne, to the Electric Boat facility in Connecticut in 1949. He created his home in Deep River, CT. Chase died on May 9,1974 at the age of 90. (Irwin Chase)*

Figure 1-3. *Glenville S. Tremaine had been a Ship Fitter's apprentice at the Bath Iron Works in his native Maine, and he took correspondence courses in draftsmanship at the same time. He studied yacht design under Henry Douglas Bacon and was "loaned" to naval architect Morris Whitaker for a brief assignment to New York in 1912. When work at Bath slowed down, Tremaine answered a newspaper ad for a draftsman and was hired by Elco in March, 1913. Tremaine became Irwin Chase's apprentice and assistant draftsman until the 1923. Tremaine carried on the Elco design traditions that Chase began in 1906, and he drew wonderful cruisers and motor yachts through the Roaring Twenties, aided by his new assistant, Alfred "Bill" Fleming. It was Tremaine that padlocked the doors, to end Elco's long history of wooden boat building in 1949. He would go on to open his own company building boats called Tremaine Yacht Works in Ship Bottom New Jersey. (Glenville Tremaine).*

Figure 1-4. *Preston L. Sutphen was the son of Elco's Executive Vice President, Henry R. Sutphen. He was an Army pilot in World War I and would attended Princeton University, graduate of the class of 1921. He would begin his employment with Electric Boat Company in 1922, at it's Groton plant, before transferring to Bayonne in 1929. He worked in the Sales Department before becoming General Manager of both the Yacht Division and Elco Naval Division. During the war Sutphen visited forward bases to see how the PT boats were holding up in the Pacific and recorded ways to make things better. His son Preston Sutphen II would serve as an officer on PT-563 with Squadron 29 during the war. Later after the war it was Sutphen who competed with other boat builders in the pleasure boating field. After his retirement he served from 1952 to 1962 as a consultant on small boats for the Government. (Preston Sutphen)*

Figure 1-5. *Lt. Cmdr. Frank W. Rowe was the Assistant Supervisor of Ship Building for the Elco Naval Division. In addition to his regular duties Rowe initiated, organized, established, and in 1940 began operating a school behind the plant for familiarization and primary operational training of the officers crews for the first PT boat squadrons. When the Motor Torpedo Boat Squadrons Training Center was authorized in March 1942, the Elco Naval Academy was then transformed into a school of specialized instruction in the construction, maintenance, and repair of the PT boats, and would continue until the establishment of the MTB repair and training unit at Melville, Rhode Island. By December of 1944, in excess of 2500 officers and enlisted men were given instruction during the four years the Elco Naval Academy operated. (Heather Rowe)*

Figure 1-6. *Alfred "Bill" Fleming came onboard Elco around 1923. He would become Glenville Tremaine's assistant, and shortly would become Elco's premiere Stylist. Fleming was a pure behind the scenes genius and a real stickler for details when it came to the design of pleasure craft in the roaring twenties. His work ethic and attention to the most smallest details served him well when the PT boats were being constructed. (Bert Morgan)*

Figure 1-7. *Building 21, the main assembly building for the PT boats. Still under construction, she was designed and built by the Austin Company of New York. At the time she was a state of the art wooden boat building facility. Forms for the concrete roadway will be the next phase of the project. The two large doors were designed to slide open, so boats could be towed onto the concrete roadway. (Irwin Chase)*

Figure 1-8. *The backside of the building was designated as Building 20, where the main offices would be located. An area was set aside for the US Navy and Supervisor of Shipbuilding. Tons of concrete will be mixed and poured before the walkways and roadways would be completed. To the left of this photo we see the two massive outdoor lumber sheds. (Irwin Chase)*

Figure 1-9. *Construction has begun on what would become the large outdoor boat shed. A barge crane will lift the heavy steel girders into place for assembly. When completed this area will hold several boats in dock spaces where other areas of assembly can take place. This shed will alleviate space problems in the main plant area, and give the boats time to soak. Boats running time trials or being used for other testing will operate from this shed area. (Irwin Chase)*

Figure 1-10. *Building 21 construction continues to move forward, and was completed in an amazing time of 100 days. The concrete roadway is completed and work continues on the massive twin boom crane. This crane was built and assembled by Bergen Point Iron Works in New Jersey. The Bergen Point Iron Works manufactured steel structures and highway safety devices. The company began business in February 1907; its first job was a small extension to the Narragansett Bay Coal Depot. The firm went on to construct naval coaling stations at Cristobal and Balboa on the Panama Canal. To the left we see the completed outdoor boat shed, and docking facilities. (Irwin Chase)*

Figure 1-11. *Taken on March 16, 1940, we see an inside look at Building 21, the main assembly building before she would become crowded with boats. Equipped with two main assembly aisles and companion aisles for sub assembly work, this plant provided for several production lines, in which each boat moved forward on its own individul cradle. This area was a massive 240×420 and also included an upper mezzanine level, outdoor boat shed and attached Elco offices. (Irwin Chase)*

Figure 1-12. *Long view of Building 21, the main assembly area. This looks forward toward the doors which lead out to the concrete roadway and dock area. The boats would make their way to the concrete roadway for launching, for another phase of assembly and soaking. (Irwin Chase)*

Figure 1-13. *Work continues on what would become Building 16. This area would become the wet boat basin, where boats would be towed to complete and free up space in the main assembly building of Elco. At least 6 boats would fit into this area at one time. This building was not completed in time to be used however on the seventy foot designed boats. (Irwin Chase)*

Figure 1-14. *Much thought went into the design of how the main assembly area would be set up. Here we see one of the machine shops on the construction floor. It was close to the main area where boats were being assembled, with no steps lost in providing equipment as quickly as needed. (Irwin Chase)*

Figure 1-15. *On February 10, 1939 Henry R. Sutphen Executive Vice President of Elco, and at the time Chief designer Irwin Chase, embarked for England on the Queen Mary. They would check out the information on the Royal Navy's progress with MTB designs, by observing the types under construction and see for themselves performance tests. They believed that small, seaworthy motor boats, capable of high speeds would make for a useful fighting craft. Preliminary talks resulted in an understanding that if they succeeded in purchasing one of the 70 foot Scott-Paine designed boats, the Navy would purchase in accordance with the terms of a contract tentatively agreed upon by Elco and the Navy. Negotiations for the Scott-Paine designed craft were successful and on March 17, 1939 Sutphen reported by radio that the boat was purchased under favorable terms with shipment to America in three months. This photo shows Sutphen (4th from Left) with Chase to his left, on the RMS Queen Mary. (Henry R. Sutphen)*

Figure 1-16. *The Scott-Paine boat PT-9, has made her way from England as deck cargo aboard the SS President Roosevelt, arriving on September 5, 1939. Here she is being off-loaded in New York. From here she was lightered to the Electric Boat Company, in Groton, Connecticut. Elco moved its investment to New London for a series of tests. (Irwin Chase)*

Figure 1-17. *A nice starboard side look at PT-9 during testing. Scott-Paine himself came over with the boat, and was in charge of its operation. He, together with the Elco company, solicited the Navy to send its representatives up to New London, to take a spin. The Coast Guard who was also interested sent an observer to these trials, although they did not think this boat had the potentiality for a Coast Guard boat, feeling it was too limited. (Irwin Chase)*

Figure 1-18. *Port-side look at PT-9 as she continues her testing in New London. One can clearly see her flared bow and streamlined look. You also get a good look at the Dewandre Power gun turrets. At the helm is Elco's Managing Constructor Irwin Chase. (Irwin Chase)*

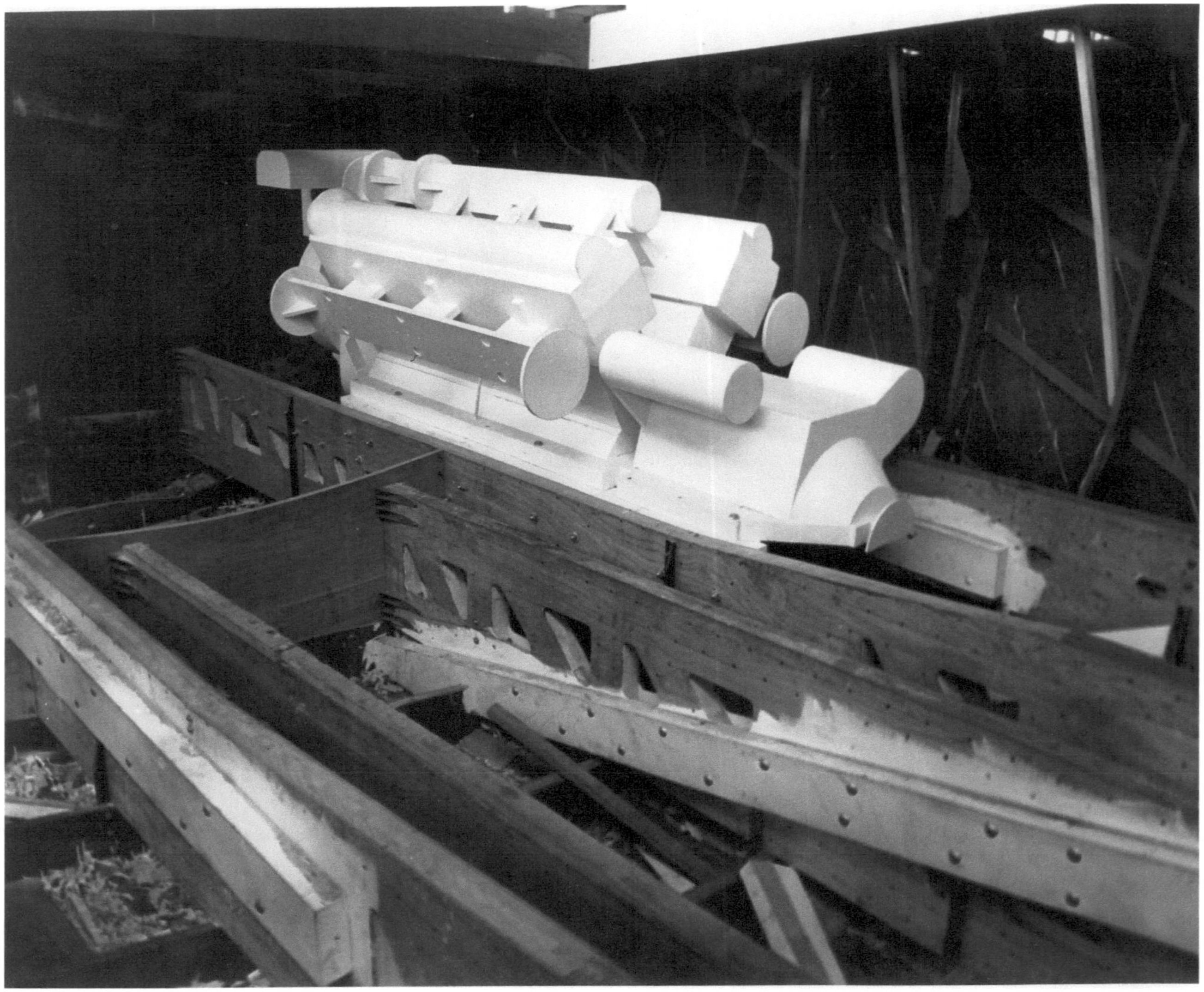

Figure 1-19. *At the time PT-9 had arrived it was powered by three Rolls Royce Merlin engines. It was felt by Packard chief designer J.G. Vincent, that using an American engine could better the performance of the Rolls Royce product. He convinced Elco and PT-9 would later have Packard Marine Engines installed. Here mock Packard engines are being placed into the engine beds for proper placement. (Irwin Chase)*

Figure 1-20. *Engineers go through their paces in observing all aspects of the Scott-Paine design PT-9. The Government acquisition of the boat was to take place on November 1, 1939, but the delivery date would be postponed until January 1940. In reality the delay was almost six months but it gave the Elco company time to study the construction of the boat as they were prepared to build their own version of the Scott-Paine design. Elco would base its first series of boats upon this Scott-Paine model, and the delivery delay gave engineers an ample opportunity to examine the boat, look into its construction principles, and adapt or modify the design characteristics of the English boat to best suit the needs of our Navy. (Frank J. Andruss Sr.)*

Figure 1-21. *With Elco given the green light by the trial board in successful tests of PT-9, the Navy ultimately approved the purchase of PT-9. The Navy Department made an award to Elco to build 10 motor torpedo boats, PT's 10-20 and 12 motor boat sub chasers PTC's 1-12. The first step in the construction of the boat was the assembly of the prefabricated main frames on a building jig which is a double 12- by 12-inch wooden base extending the entire length of the boat. The frames are built of wood braced with plywood. This framework will undergo many assembly stages before completion. (Irwin Chase)*

Figure 1-22. *The jig with its many frames has been completed and now the keel been added. Workers have started to assemble batten strips on the boats bottom, and sides. The spruce keel, to which is spliced to an oak stem, is put in place and secured with brass bolts. The chine, which is the hull member forming the joint between the bottom of the hull and the sides, is made of spruce and also bolted on. Spruce diagonal frames on the hull sides and longitudinal battens on the bottom are notched into the main frames and fastened with brass screws. Intermediate frames between the main frames are similarly installed in the three forward compartments. The hull is now ready for planking. We can also see here the Elco innovation of building the hull from bottom up, making it easier for workers to work from a less cramped position. (Irwin Chase)*

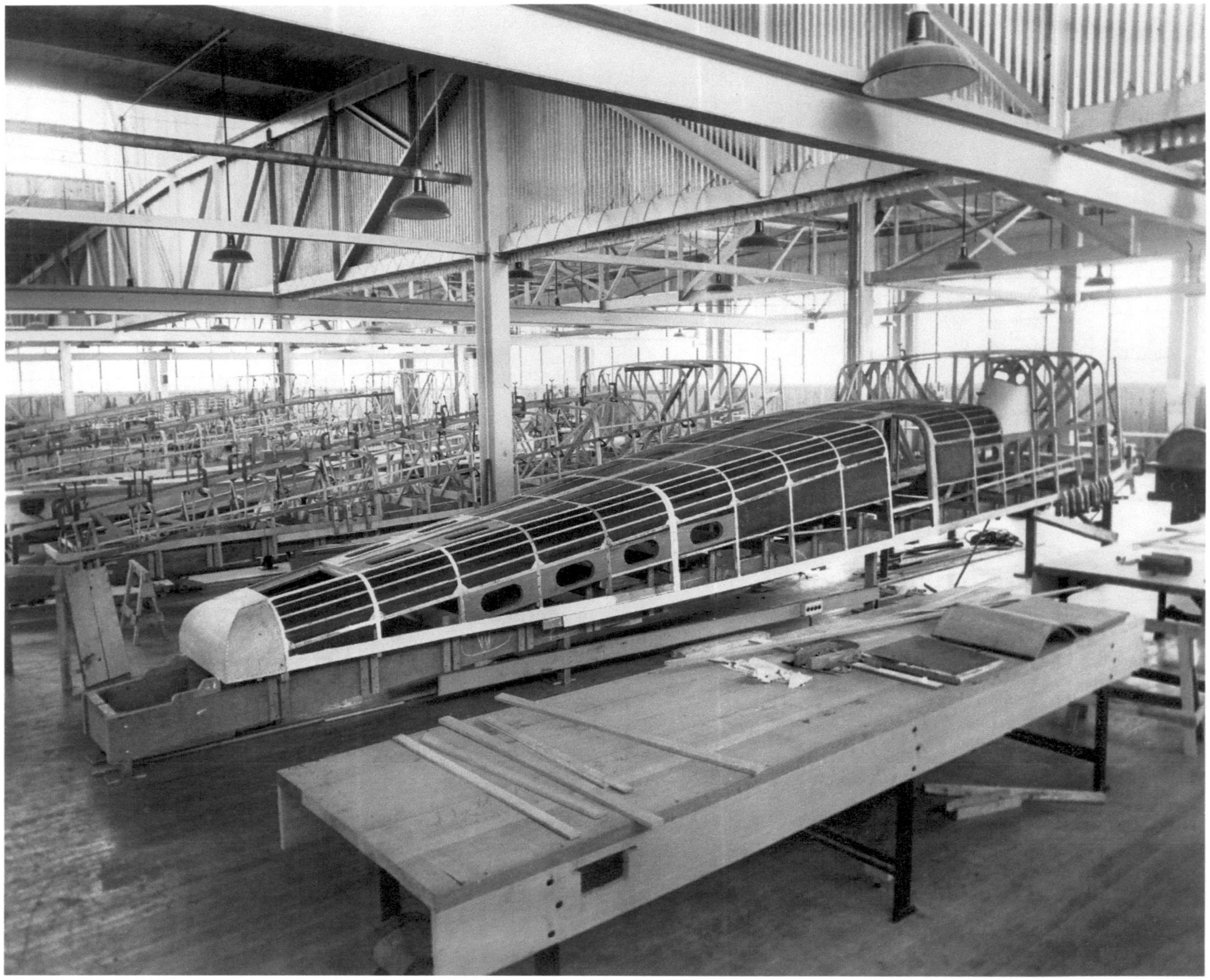

Figure 1-23. *A look at the joiner shop located in the upper mezzanine of the main assembly area. It was ideal for making structures and other sub assemblies. Here a rear cabin trunk structure is being framed out for one of the 70 footers. You can clearly see the light weight construction that went into the making of these structures. When completed the cabin was secured to cables and hoisted to the first floor assembly area, then mounted onto the hull. (Irwin Chase)*

Figure 1-24. *Another look at the joiner shop, showing progress being made on the cabin trunk structures. Wood working machines and tools were kept here as all assembly of these structures was done in this area for the 70 foot designed boats. (Irwin Chase)*

Figure 1-25. *When planking is complete, the hull is turned over and the inside work begun. Longitudinal girders made of mahogany covered with plywood run from bulkheads 1, 2, and 3 back to the stern and give the boat a great deal of its stiffness. Cedar clamps or battens are screwed to the inside of the frames on the sides of the hull and run from bulkhead No.1 to the stern. The main deck frames are notched, spruce deck longitudinal are screwed in place and a 5/8-inch mahogany plywood deck is put on. The deck is secured to the hull with plywood knees between each frame. The knees are bolted to the deck and to the covering board on top, and to the planking, gunwale, and guard clamp on the sides. (Irwin Chase)*

Figure 1-26. *A wonderful look at one of the boats looking forward. On the bow her tow rails have been added, as well as the boats forward cabin structure. We see the deck filler on the port side decking which has yet to be sanded. Once sanded a non slip paint will be added to the deck. Workers are below deck doing wiring and piping assembly. On the port side mechanical engineer Clyde Anderson is looking into the cabin structure, keeping a keen eye on all that is being assembled. There is no room for any errors in the construction of the boats. (Irwin Chase)*

Figure 1-27. *We see the seventy-footer's in different stages of assembly. Notice the three boats from left to right have had their alunimum alloy boots installed. They were placed on the front of the pilot house just below the three cabin windows. This space was created to stow gear. To the rear of the assemly area we see one of the boats which has been numbered PT-17. She has her torpedo tubes installed as do the other boats to her left. (Irwin Chase)*

Figure 1-28. *In the assembly building we see several of the 70 foot boats in different stages of assembly. Most of the boats have had their hulls completed, and at this phase of assembly we see the boats with and without decking, while the boat in the center has it's planking in place. The forward cabin has been installed, and awaiting workers to continue the process of adding toe rails, hatches, and other equipment. Some of the more obvious changes from PT-9 were in the boats superstructure. The canopy was enlarged and it was constructed so that it could be removable. The pilot house was enlarged and flushed with the truck canopy, which helped with wind and spray deflection on the open bridge. A space was created to stow gear by fitting an alloy boot to the front of the pilot house below the windows. (Irwin Chase)*

Figure 1-29. *A nice look at the soon to be installed compressor. This large compressor will provide power to the Dewandre machine gun turrets on the boat. This will provide compressed air at relatively low pressure, needing small pipes and fittings. The air transmits pressure using oil as the hydraulic medium. (PT Boats Inc.)*

Figure 1-30. *A look at the bubble mount for the Dewandre powered machine gun turrets. These turrets really did not find favor for several reasons, one being they carried a small magazine for rounds, changing the magazine was too slow and cramped, speed of the turret was too slow, and the air compressor was found to be too small causing the turret not to operate smoothly. (Irwin Chase)*

Figure 1-31. *A small compact engine room showing the three Packard Marine Engines. Notice the levers which the engineer uses to put the engines into gear as needed. From neutral, forward, and reverse he gets those instructions from the Captain at the helm. This is looking aft where you can see the exhaust piping (white) that runs out of the transom. You are seeing the engines from the reverse gear end with the carburetor on the wing engines facing forward, and the center engine facing aft. (Irwin Chase)*

Figure 1-32. *Two Elco workers are installing the Starboard Propeller Shaft through the propeller strut, through the cut-less bearing, into the shaft log, and finally into the engine room. Once the other end of the shaft is inside the engine room, the keyed coupling flange can be installed onto the end of the shaft. This whole procedure takes about 4 people to man-handle the heavy monel prop shaft into the correct position and involves using a lot of lubrication (grease or tallow) and then twisting, pulling, and pushing! Only then can the shaft and coupling alignment process begin. It takes several hours to complete the process for three shafts. (PT Boats Inc.)*

Figure 1-33. *PT-10 begins her journey from Building 21, stern first to the concrete roadway. She is being pulled by two small tugs, and sits in her boat cradle on wheels. From here she will make her way to the dockside crane, and gently placed into the boat basin. (Irwin Chase)*

Figure 1-34. *Clearing the large sliding doors of Building 21 took skill and patience. At this point she needs to complete other assembly phases, as she carries no machine guns, torpedo tubes, and other important pieces of equipment. Soon the Elco test team will fire up her engines, while she sits at the dock. (Irwin Chase)*

Figure 1-35. *PT-10 has been brought along the concrete roadway and has been hooked in the dockside crane. She will be gently lowered into the boat basin and then towed to the next phase of assembly. The boat sits in the boat cradle which was fabricated and furnished by Savary & Glaeser in Dunellen, New Jersey. As you can see she has yet to have her machine guns installed, torpedoes, ventilators, and other top side construction. The boat typically needed to "soak" or remain in water for several days to allow the wood to fully absorb moisture and expand, creating a tight seal between the planks and preventing leaks; the exact time depended on the wood type, boat design, and water conditions (Irwin Chase)*

Figure 1-36. *Dignataries and other Elco staff stand on the deck of PT-10, shortly after being placed into the boat basin. Here she has been towed to the outdoor covered boat shed in front of Building 21. Sub assemblies will take place here, and yet again she will be towed for dock side engine warm up for testing. (Irwin Chase)*

Figure 1-37. *A look at PT-10 in all her glory, as she is running just outside of the Elco Naval Division. The boat is still in Elco hands and will be put through several tests and trials before being placed with Squadron Two. She would be under the command of Lt. Cmdr. Earl S. Caldwell, USN, where she would accompany her sister boats on a shakedown run in Florida and Caribbean waters. She would not serve with the US Navy but was transferred to the Royal Navy on April 11, 1941. (Irwin Chase)*

Figure 1-38. *Another beautiful look at PT-10, as she continues her training run. One can clearly see the wake created by the powerful Packard Marine Engines, and the design of her hull. This was something that was a constant problem for the boats and their crews, as these wakes could be seen from the air by enemy planes. (Irwin Chase)*

Figure 1-39. *Taken at the New York Navy Yard in Brooklyn, these newly built PT's 9-12 along with PT's 13 and 14, were delivered here. At the time Elco was rushing to deliver the remaining boats, PT's 15 through 19. To the left of the photo is PT-3 built by the Fisher Boat Works, Detroit, MI. This photo taken around November 1940, shows the boats shortly before they would be ordered on December 16th, to proceed to Key West, Florida. They would also cruise from Key West to Guantanamo, San Juan, Puerto Rico, and St. Thomas in the Virgin Islands. This would become the boats shakedown cruise, where they would remain in warm waters until during the cold months, were to return on May 13, 1941. (Frank J Andruss Sr.)*

Figure 1-40. *A look at the Elco 70 foot boats while they were at the Washington Navy Yard in Washington D.C. around December 1940. Ahead of the boats are the USS Potomac (AG-25) and Cuyahoga (AG-26). Some of the crew from these boats are on the dock, taking a much needed rest before beginning the next leg of their journey. From here they will proceed to Key West Florida, to begin their shakedown cruise to Cuba. (Naval History and Heritage Command)*

Figure 1-41. *A wonderful look at PTC-1, one of the first Submarine Chasers built according to the contract. Here in her cradle, she is hooked into the dock side crane and awaits launching into the boat basin at Elco. She will begin her trails soon after launching. (PT Boats Inc.)*

Figure 1-42. *After being launched, we see PTC-1 on her trials. Notice they were identical to the 70 foot PT boats, except they carried depth charge racks and Y-guns to combat submarines. This boat was placed in service on March 6,1941 and assigned to Motor Boat Submarine Chaser Squadron One, under the command of LT (jg.) John D. Bulkeley. This idea was abandoned because of failure to develop satisfactory sound gear to enable the boats to locate submarines. (Frank J. Andruss Sr.)*

Figure 1-43. *Another PTC showing the unique depth-charge system in greater detail. Y-gun arbors were stowed alongside the superstructure, and charges moving aft along the rails were hoisted up the inclined racks. Her Dewandre Turrets are covered with canvas. (Frank J. Andruss Sr.)*

Figure 1-44. *A look into the upper helm area of one of the PTC boats. As you can see the helm is identical to the PT boats. The control box to the left contains 8 plunger buttons which operate the launching of the depth charges using a solenoid. The small box is the launching key which must be switched on to provide power. Notice the plunger buttons have a safety strap over them, to prevent accidental arming. (Frank J Andruss Sr.)*

Figure 1-45. *A look at the inside of one of the boats Dewandre Turrets without the enclosed bubble. You can see some of the pipes and fittings that help operate this turret. The motive power for operating the turret was compressed air at relatively low pressure. An air compressor which was electrically driven, had the capacity of 10 cubic feet of free air per minute supplies air to a flask, of 1 cubic foot capacity at maximum pressure of about 240 pounds per square inch. When the air pressure in the flask falls to 150 pounds per square inch, the air compressor automatically starts and recharges it. As you can see, there was not much room for a gunner in here and very small ammo capacity boxes. (Irwin Chase)*

Figure 1-46. *One of the PTC 70 footers which were numbered 1-12, races up Naragansett Bay, after passing the Mt. Hope Bridge. []{#anchor}Opened in October 1929, the bridge is located at the mouth of the Taunton River, on the border of Massachusetts and Rhode Island. It connects the Rhode Island towns of Bristol and Portsmouth. The boat is heading to Newport to join her sisters and the tender USS Niagara, which at the time was set up as a familiarization school aboard the tender for prospective PT personnel. These boats were transferred to Great Britain in the spring of 1941.(Naval War College Archives)*

Figure 1-47. *A rare look at a color photo of of the Elco 70 footers. On this day they are docked in Florida. With winters in New York hampering operations, the boats would head to Florida with their scheduled base being in Key West. They would remain here in the cold weather and return to New York in May 1941. (Kingston Winget)*

Figure 1-48. *Another color photo showing the boats in Key West Florida. Their Dewandre turrets are covered with canvas as well as the aft torpedo tubes. Duty on the boats here was much more conducive to underway training, and from the looks of things a pleasant area to be stationed. (Kingston Winget)*

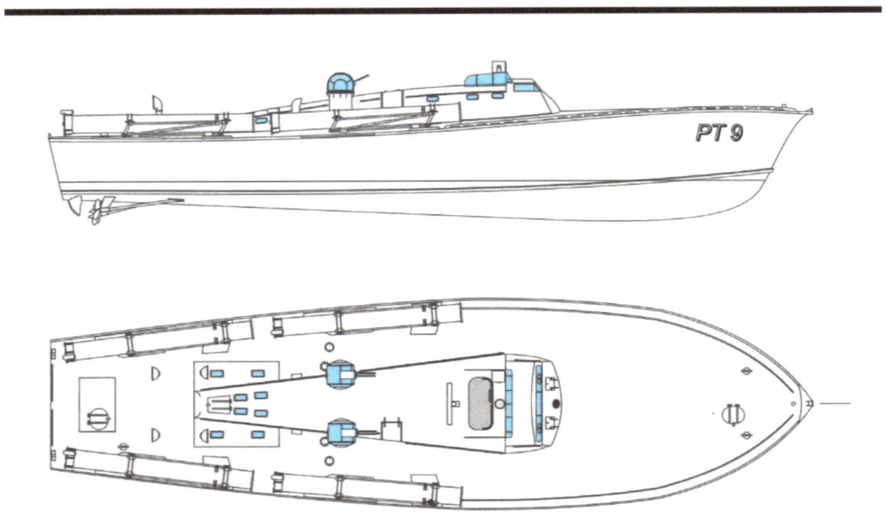

Figure 1-49. *This is a line drawing of PT-9 showing her starboard side and overhead look as was designed by Scott-Paine and constructed by the British Power Boat Company. This was the boat that would be shipped back to Elco, and was the starting point for the Elco 70 foot PT boats. (Drawing by Al Ross)*

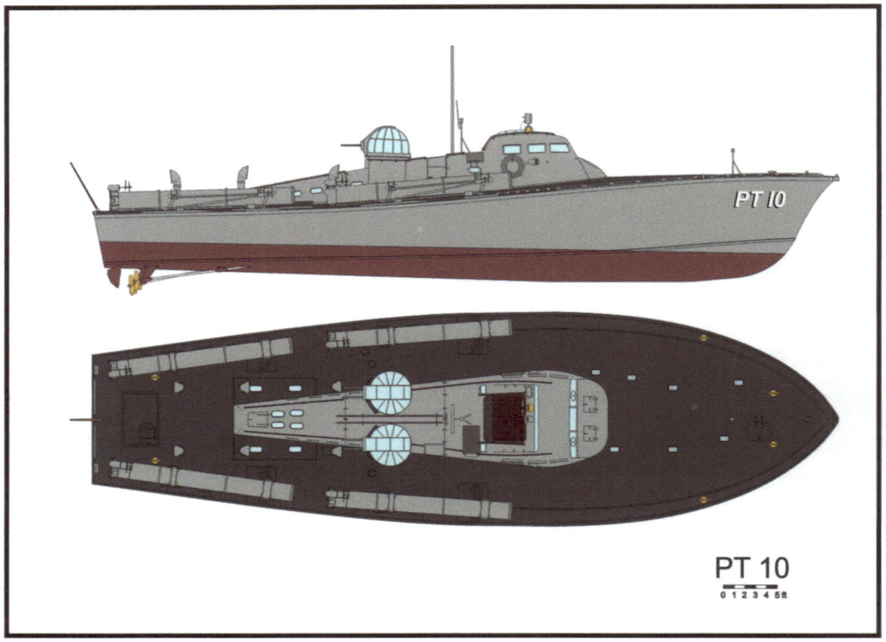

Figure 1-50. *Beautiful artist rendition of the first of Elco's 70 foot designed boats. PT-10 as she was built and painted coming out of the factory. A nice look at her starboard side and her deck arrangements from over-head. (Drawing by Al Ross)*

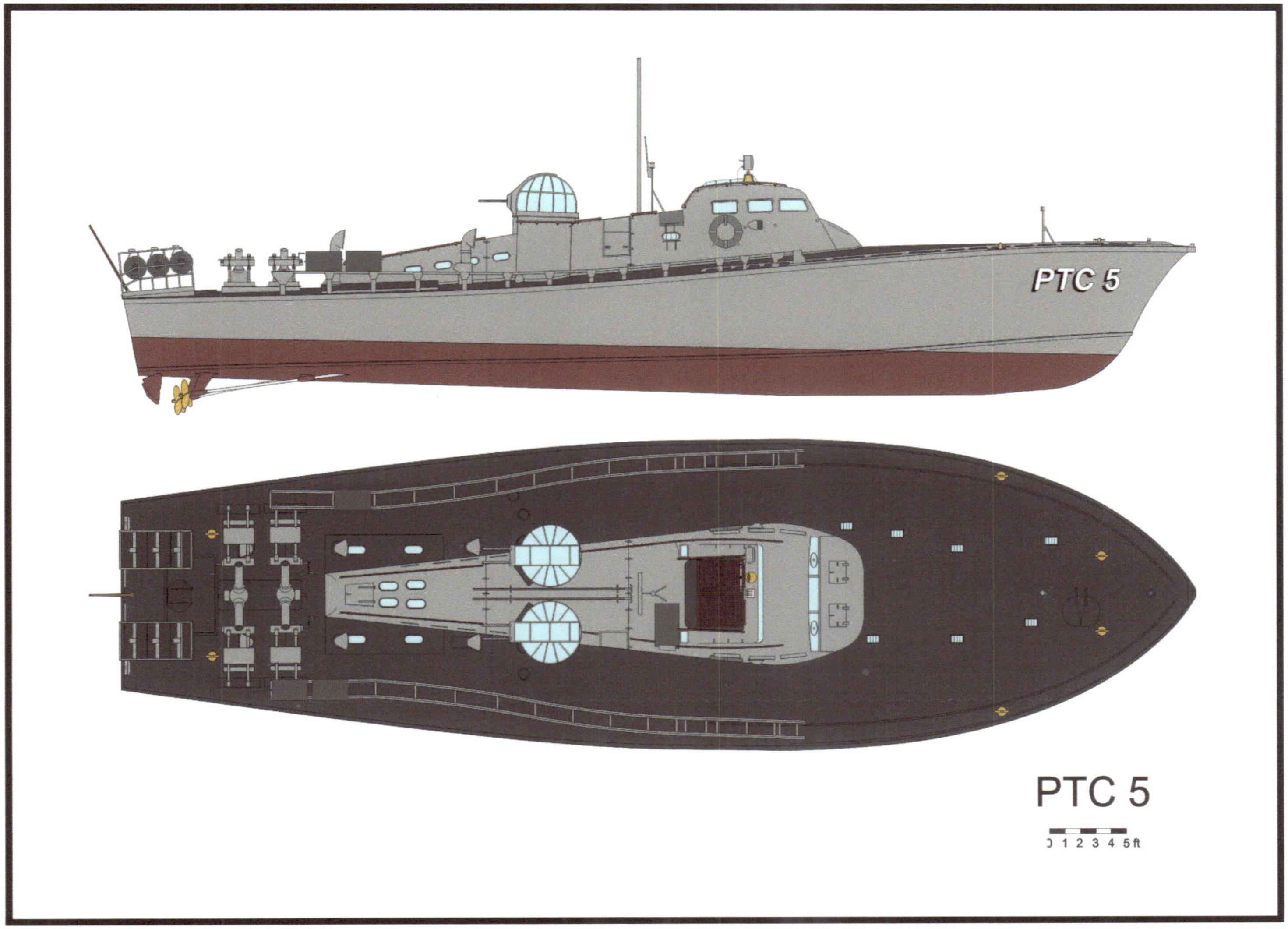

Figure 1-51. *This drawing shows one of the PTC submarine chaser boats built by Elco. Identical to the PT boats 10-19, these were set up for submarine duty. Starboard side and topside view show these boats as they came out of the factory, complete with depth charge racks and Y-guns. (Drawing by Al Ross)*

Chapter 2

Elco's Seventy-Seven-Foot Design

Chapter Two

It was due to design completion plans that the Bureau of Ordnance decided that the larger boats would have the twenty-one-inch torpedo tubes installed, which would mean an increase in the size of the boats.

Before any changes could be made the bureau decided to review its data on the PT boats and obtain new opinions. The board actually recommended an eighty-foot wooden boat, that was powered by reliable gasoline engines, capable of forty knots with a range of at least five hundred miles at twenty knots. Armament would consist of four twenty-one-inch torpedoes and tubes, one smoke generator, four .50 caliber machine guns in two twin mounts, with five thousand rounds of ammo per barrel. The board also required a satisfactory radio for operational use, and provision stowage for five days.

The Navy would sign a second contract with Elco on September 17, 1940, which was slightly modified on September 25, 1940. This contract called for the construction of PT's 21–32 and PTCs 33–44. They would be seventy-seven feet overall. The PTC clause of the contract was never filled, due to the unsuccessful experiments with the earlier type that convinced the Navy that other craft would be more successful as antisubmarine vessels, so the PTC boats 33–44 were built with the same numbers.

The Elco PT-20 class boat was basically a boat with seven feet added on, but it was more than that. The reverse sheer of the hull was not as pronounced as it had been on the seventy footers, and the distance between the cowl vents on the stern also showed the new length to good advantage. Considerable strengthening of the hull could be seen below decks and the intermediate framing was added in the forward section of the pounding area of the bow. Eighteen inches would be added to the boats engine room, and the lazarette was enlarged by five and a half feet, this totaling seven feet. The boats were a V-bottom, hard chine design with a draft ranging from two and a half feet forward to four and a half feet aft, and have a loaded displacement of approximately forty-six tons. A hard chine design has very little V-shape on its bottom whereas a soft chine boat has a considerable V which gives a sharper entrance into the seas. The hard chine design gives greater speed due to the fact that it causes the boat to plane.

The hard chine design gave greater speed, due to the fact that it causes the boat to plane, but it also causes the boats to become extremely wet in choppy or very rough weather, particularly at slow speeds. The first step in the construction of the boats was the assembly of the prefabricated main frames on a building jig which was a double twelve-by-twelve-inch wooden base extending the entire length of the boat. The frames were built of wood, braced with plywood angles. The bottoms of frames one to thirty-four were laminated of spruce, white oak, and mahogany; the upper part of frames one to thirty-four and the entire frame from thirty-five to sixty-nine were made of mahogany. All the frame joints were secured with marine glue and screws. Bulkheads are merely frames, properly braced and covered on both sides with fir plywood. They are, for all practical purposes, watertight.

The spruce keel, to which is spliced to an oak stem, is put in place and secured with brass bolts. The chine, which is the hull member forming the joint between the bottom of the hull and the sides, is made of spruce and also bolted on. Spruce diagonal frames on the hull sides and longitudinal battens on the bottom are notched into the main frames and fastened with brass screws. Intermediate frames between the main frames are similarly installed in the three forward compartments. The hull at this point was ready for planking.

The planking is in two layers both made of mahogany boards six inches wide. Both layers above the chine are three-eighths inch thick; on the bottom, the inner layer is three-eighths inch, the outer layer seven-sixteenths inch. The two layers are laid on diagonally opposite to each other with sheets of airplane fabric laid in marine glue in-between. The planks were fastened to the main frames below the waterline with monel screws and above the waterline with brass screws. The fastenings to the diagonals, battens, intermediate frames and securing the planks to each other where there are no frames are copper nails riveted over burrs. When planking was completed, the hull would be turned over and the inside work begun. Longitudinal girders made of mahogany covered with plywood ran from bulkheads one, two, and three back to the stern which give the boat a great deal of its stiffness. Cedar clamps or battens were screwed to the inside of the frames on the sides of the hull and ran from bulkhead No. 1 to the stern. The main deck frames are notched, spruce deck longitudinals are screwed in place and a five-eighths-inch mahogany plywood deck is put on. The deck is secured to the hull with plywood knees between each frame. The knees are bolted to the deck and to the covering board on top, and to the planking, gunwale, and guard clamp on the sides. An oak rub strake outside the gunwale and an oak chine guard make the outer hull work complete.

The gasoline tanks were mounted on padded frames and secured in place with padded metal straps. Each tank is provided with its own separate fill pipe and each one vents immediately below the gunwale on the outside of the hull. The engines are mounted on bed frames which are secured to the longitudinal girder. Shims were provided for aligning the engines properly.

The canopy was framed with spruce with all corners and angle joints made of mahogany and fastened with screws and glue. The frame is covered, inside and out, with light plywood which is secured with glue. The voids between frames are filled with rubber insulating material. For convenience in removal of tanks and engines, the canopy is made in three sections, each of which can be removed. The forward or pilot-house section contains the sheet aluminum cockpit and the controls. The midships section contains the gun turrets which were plywood except for a thin metal shield on the top. The after, or engine-room section, is very easily removable and is fitted with a hatch large enough to permit the passage of a reverse gear, V-drive or auxiliary generator. All canopy joints are covered with tape made of airplane fabric to make them watertight and metal clamps are used to hold the two sections together.

A great amount of deck cracking was experienced in the seventy-seven-foot boat due to the fact that the deck did not have as much flexibility as the hull. This weakness was eliminated by the installation of a hollow fir "strongback" which is a longitudinal stiffener on the deck from forward of the pilothouse to the lazarette hatch on both sides of the canopy. A tendency for the canopy to crack was likewise eliminated by the "strongbacks" and a slightly heavier construction and modified design of the canopy.

The boat's fire-fighting equipment consisted of several small hand CO_2 extinguishers and an automatic Lux CO_2 system. The two tanks in the Lux system were located in the starboard officer's stateroom with heat actuators in the tank room and engine room which will cause the system to discharge upon a sudden increase in temperature. A manual control was provided on the deck immediately forward of the canopy hatch.

Standard issue MK-XVII torpedo tubes were mounted on the deck, which carried the MK-VIII torpedoes. The tube was made of steel, fabricated by welding. It is pivoted on an aluminum alloy turntable and turntable base. The forward end was mounted on an aluminum alloy support which travels on corrosion-resistant steel plates and is guided by brass Z-bars, cadmium plated. The turntable and the forward support are joined with steel tubing rails. The tube did not recoil.

The VIII had many technical difficulties with its design that would be a bane to the torpedomen who would use them. The first issue came with the process of simply launching the torpedo, which was fraught with its own set of difficulties.

The gyro in the torpedo was sensitive when it was first launched, and it would need to be launched from an even keel, otherwise the torpedo would lose stability when it hit the water. The other issue concerned the launching mechanism. The initial system was set with a black powder charge to push the torpedo out of the tube. This was a problem where the humid climate would cause these charges to misfire, sometimes not putting enough force behind the torpedo to fully eject it from the tube. One other risk of these misfires was what is called "hot running", where the torpedo would run in its tube. Though the warhead could not detonate, the motor would overheat and explode without water to cool it, sending splinters across the deck. Another issue was that the tubes were heavily lubricated with oil and grease to ensure that the torpedo was smoothly launched from the tube. Sometimes during launch, the charge would ignite the lubricants in the tube causing a fire emitting black smoke that would reveal the location of the PT boat.

The power plant of the Elco seventy-seven-foot boat consisted of three 1,350-horsepower Super-charged Packard 4M-2500 gasoline engines. The cooling water and lubricating oil in the engines were cooled by use of a heat exchanger in which the cooling medium is seawater. The heat exchanger is a metal cylinder divided into two sections with fifty to seventy-five one-quarter-inch outside diameter copper tubes inside through which salt water flows. The oil and water flow over the tubes, each in its separate section, and are cooled by a constant flow of seawater.

As the engines are installed in the boat, the two wing engines were installed backwards, the reverse gear being towards the bow of the boat. To drive the propeller shafts which run aft, a gear box called a V drive was used. The V drive consists of two cone-shaped gears with roller bearings at each journal. The engine drives one gear which drives the other which, in turn, drives the propeller. Cooling is accomplished by use of water jackets through which seawater is pumped by the engine salt-water pump. A separate oil tank for each V drive was located in the bilge under the unit. Oil is put in the housing of the V drive itself and its level is determined by a bayonet type gauge on the filler plug. Lubrication was furnished by an oil pump driven by the unit itself.

The two-inch Monel propeller shafts go through the hull through a unit called the stern tube. The stern tube is a bronze casting with a flax-packed gland on the inboard end which keeps the water out of the hull. A packing nut tightens or loosens the packing. Cooling is obtained by circulating salt water with a small scoop. A bronze bushing is lubricated by a permanently installed grease gun. The three propellers were bronze and keyed to the shaft with a bronze key. Great care must be exercised in the installation of propellers to be certain that they are all the way "home" on the taper and not riding the key or off center.

The boats gun turret's in the first series boats, was the Dewandre type power driven system, of the type in which a yoke carries a mount fitted for the two machine guns, which yoke is suitably pivoted an a carriage of circular form. This carriage rotated on a circular track. With reference to the carriage and track the yoke with the mount may be bodily elevated in a vertical plane. In addition to this bodily elevation the guns were capable of a certain further amount of elevation end depression in any position in which the yoke may be. The carriage moving horizontally carries the guns irrespective of their positions in the vertical plane approximately 340 degrees, and also in any position of the carriage the guns themselves have a further movement to right and left.

The mount was completely automatic. The gunner has only to manipulate the gun to aim at the target when the turret and the yoke follow all the movements of the gun sufficiently so that the aim can be kept on the target all the time. The motive power for operating the turret was compressed air at relatively low pressure needing small pipes and fittings. The air transmits pressure using oil as the hydraulic medium. These type turrets would be on PT-20–44, but would be removed beginning with PT-45 (second class) as the turret was considered too vulnerable and ineffective in combat, particularly against the fast-moving, small targets that PT boats typically engaged. The turret design was prone to fogging of the glass, making it difficult to quickly target multiple enemies. It was also felt that if the air supply was cut off, the turret would become useless.

The boats would choose the open-ring twin mounts, which improved flexibility which allowed for faster aiming and quicker target acquisition. Twin mounts could fire multiple rounds simultaneously, delivering a higher volume of fire. These first mounts were called the Colt style, a scarff-rig system. These would however be changed due to the small ammo boxes, which did not hold sufficient rounds that would be needed.

Changes to the boats turrets would come in the way of the MK-17 mounts, which provided ammunition to the guns, supplied by four boxes, two per gun. The mounts carried 250 rounds per gun. This greatly improved the mounts from the previous Colt style. The twin cradle for .50-caliber machine guns was a flexible, lightweight, scarf-ring type mount, equipped with a hydraulic system to absorb recoil shock. It was designed to be fired by either or both hands.

The second series boats, beginning with PT-46, would begin to have added to their stern, the 20MM A.A. Oerlikon cannon. The Oerlikon was primarily a close-range high-angle A. A. gun. It could be used against surface vessels, but a .50-caliber machine gun can be used with better success against landing forces, etc. Against dive bombers the 20-mm. was an excellent gun. The ammunition was supplied from a magazine that held sixty rounds.

The magazine was detachable mounted on the gun. It is filled and also serviced while removed from gun. The earlier MK-4 mount's extended from the cradle, in which the gun is fixed, to the pedestal bolted to the deck. The cradle rotates about the trunnion axis. The mount permits the height of the gun trunnion above the deck to be altered easily and rapidly for any sight angle to suit the convenience of the gunlayer. This is done by the column-raising hand wheel mounted on the pedestal head. The cannon was rated to fire at 450 rounds per minute, using high explosive shells and tracers.

Beginning with PT-46, Elco would add the early MK-II smoke generator which had two tanks. The smaller CO_2 tank was upright at the head of the Titanium Tetra chloride tank and towards the center line. This addition would prove as important to the boats survival as other weapons on the boat. When released into the air using Titanium Tetra chloride, it hydrolyzes quickly and resulted in a dense white smoke, in which the boats could hide behind and escape from enemy ships.

It was in the fall of 1941, that the Bureau of Ships, at a meeting decided the boats should be enlarged. The CNO requested that the Bu Ships write up new specifications for a new boat. Elco was invited to submit designs and then bid for the proposed boat. This would mark the end of the Elco seventy-seven-foot boats, as the new and improved 103 class eighty-foot boats would now become the principle boat that would be built at Elco.

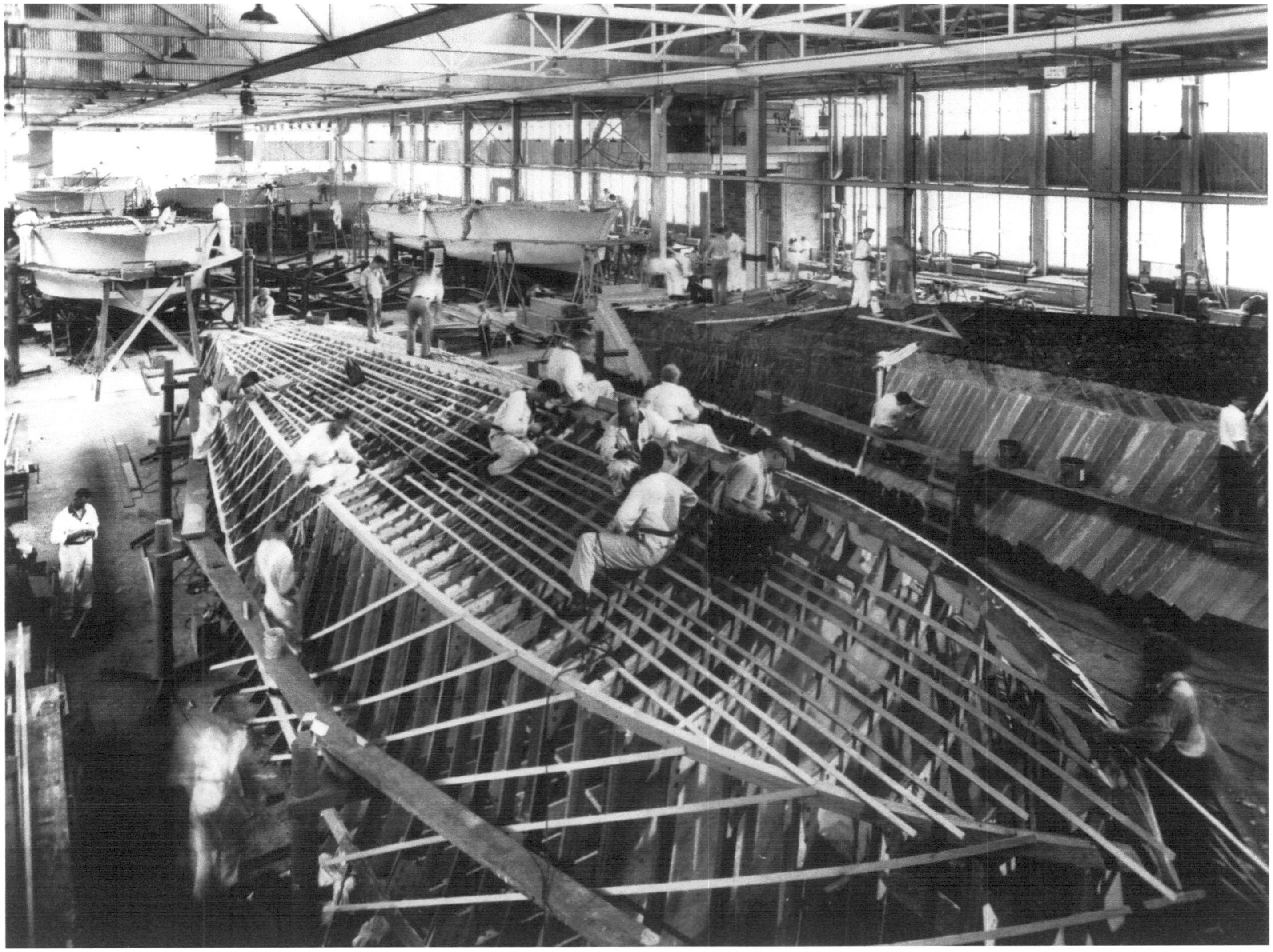

Figure 2-1. *Construction on the first series seventy-seven footers in Building 21. Workers are adding spruce diagonal frames on the hull sides and longitudinal battens on the bottom are, notched into the main frames and fastened with brass screws. Intermediate frames between the main frames are similarly installed in the three forward compartments. From these wooden strips, the first layer of Mahogany planking will be added. Workers crawl over the hull, which as you notice is built upside down, to make it easier to work on the hull. When the hull is completed it will be turned right side up. (Irwin Chase)*

Figure 2-2. *Workers are applying the first layer of Mahogany planking, now that the batton strips have been completed. The planking is in two layers both made of mahogany boards 6 inches wide. Both layers above the chine are 3/8 inch thick; on the bottom, the inner layer is 3/8 inch, the outer layer 7/16 inch. The two layers are laid on diagonally opposite to each other with sheets of airplane fabric laid in marine glue in-between. The planks are fastened to the main frames below the water line with monel screws and above the waterline with brass screws. The fastenings to the diagonals, battens, intermediate frames and securing the planks to each other where there are no frames are copper nails riveted over burrs. (Irwin Chase)*

Figure 2-3. *Inside the main assembly building looking East, we see the hulls now turned over. Work has begun on the decks and the boats are all in different phases of assembly. The main deck frames are notched, spruce deck longitudinals are screwed in place and later, a 5/8-inch mahogany plywood deck is put on. The deck is secured to the hull with plywood knees between each frame. The knees are bolted to the deck and to the covering board on top, and to the planking, gunwale, and guard clamp on the sides. An oak rub strake outside the gunwale and an oak chine guard make the outer hull work complete. (Irwin Chase)*

Figure 2-4. *A nice look into the assembly phase of these first series boats. The boat to the left has it's starboard side torpedo tubes in place, and torpedo training gear being assembled into place. Notice the boat has the hydraulically powered Plexiglas-domed Dewandre turrets, being installed on the boats at this time. The open hatch leads down into the boats engine room. (Gene Kirkland)*

Figure 2-5. *The first of the seventy-seven footers, PT-20 has been gently put into the boat basin, and sits in front of Building 21. The boat was the prototype for the first series of seventy-seven footers. She would retain several features of the seventy- footers, including the turrets and engine room canopy. PT-20 introduced decking which was a layer of mahogany plywood some 3/16 of an inch thick, used to improve stiffness. There were two series of 77 foot boats, the first being PT-20-44. The second series would be PT-45-68. This boat would be stationed at Pearl Harbor on December 7, 1941, and later stationed at Midway Island. (Irwin Chase)*

Figure 2-6. *Members of the Elco team along with members from NBC, will begin their broadcast to highlight PT-20, the first seventy-seven foot boat off the assembly line. The ceremony was broadcast to Bayonne New Jersey as well as New York. As she was the first of the Elco seventy-seven foot boats, she would soon prove her ability at the upcoming plywood derby, that she was built for speed and maneuverability. In the center of the photo is Lt. F. W. Rowe, the Assistant Supervisor of Ship Building for Elco. (Irwin Chase)*

Figure 2-7. *One of the first series 77 footers is under her own power and is returning from a trial run. She has entered the break water, heading to the Elco dock. The trial crews and other engineers ran several predetermined tests before the boats would be turned over to the Navy. Behind the boat we see the old Newark Bay Railroad Bridge, which had a vertical lift section which were a pair of two-track spans over two separate shipping channels. During the war the bridge was a critical piece of the Port of New York and New Jersey's logistic network, allowing 300 or more trains per day to run. (Irwin Chase)*

Figure 2-8. *PT-30 is lined up for the first nicknamed "Plywood Derby" on July 24, 1941. It was set up by the Board of Inspection. They would run this test to observe not only the speed of the boats, but the sea keeping and structural qualities. Elco would enter PT's 20, 30, 31, and 33. The derby was scheduled for waters off New London, Connecticut, and the course would be some 190 miles at full throttle in open sea. Elco would capture top honors in the sea run as PT-20 crossed the finish line first with a 39.72 knot average. PT-31 was second insuring an Elco victory, with 37.01 knots. Other tests such as maneuverability tests and running with a full war load were completed. (Irwin Chase)*

Figure 2-9. *Here we see a photo with work on the second series hull taking shape, as the frames have been put in place. The frames are built of wood braced with plywood angles. A cabin truck structure is in the air, suspended by the pulley system, after completion in the upper mezzanine area. When completed they were lowered to the assembly room floor for attachment to the boats. (Irwin Chase)*

Figure 2-10. *In the main Building 21 area, we see the second series seventy-seven foot boats in varying degrees of hull assembly. Workers are installing the baton strips on the hull with which the planked mahogany will be attached. To the left is the machine shop where sub assemblies took place. These assemblies could then be easily brought to the assembly floor. (Irwin Chase)*

Figure 2-11. *Workers on the assembly room are working inside the boats forward canopy structure. The canopy is framed with spruce with all corners and angle joints made of mahogany and fastened with screws and glue. The frame is covered, inside and out, with light plywood which is secured with glue. The voids between frames are filled with rubber insulating material. (Irwin Chase)*

Figure 2-12. *Three of the second series boats in the assembly building. To the left PT-59, PT-58, and PT-67. Notice PT-59 has installed a loop on the top of the cabin. This was an R.D.F. or Radio Direction Finder. As the boats were small and operated at night, the sources of signals would be radio beacon stations or light vessels. The RDF however did not find favor as it was constantly needing attention to the calibrations, and the magnetic forces which existed on the boats had a disturbing effect on incoming signals. This would not find favor and would be removed from the boats. (Gene Kirkland)*

Figure 2-13. *A wonderful look at two of the second series boats. One boat has been hooked into the dockside crane, and shortly will be gently placed into the boat basin. The other boat has been brought over to the concrete roadway and awaits here turn. Both boats have much sub assemblies that need to be completed and will be towed over to the wet boat basin. Another boat is inside the outdoor boat shed to the left of this photo. (Irwin Chase)*

Figure 2-14. *A look inside the port side pilot house of PT-59. In this area would be the boats quartermaster that was in charge of all charts. He would plot courses here, and maintain the boats radio communications. He was in charge of all call signs and maintained the compass, and RDF equipment. Semaphore equipment would be kept here including the boats binoculars. (Gene Kirkland)*

Figure 2-15. *Starboard side looking forward inside the pilot house of PT-47. Take note that the photo was taken in 1945, so the place is not too ship shape, with equipment already having been removed. The door to the right would lead to the forward crews quarters. Notice the control panel just under the counter, which was duplicate controls from the main bridge. (Gene Kirkland)*

Figure 2-16. *Inside the pilot house we see the charts and chart table. From the looks of things this officer is very busy trying to plot a course. At the time, the boat was in Martha's Vineyard and the Block Island area, and was with the Motor Torpedo Boat Squadrons Training Center. Usually a list of call signs and other information that would be useful to the Captain and the boat, was kept in this location. (Charles B. Jones)*

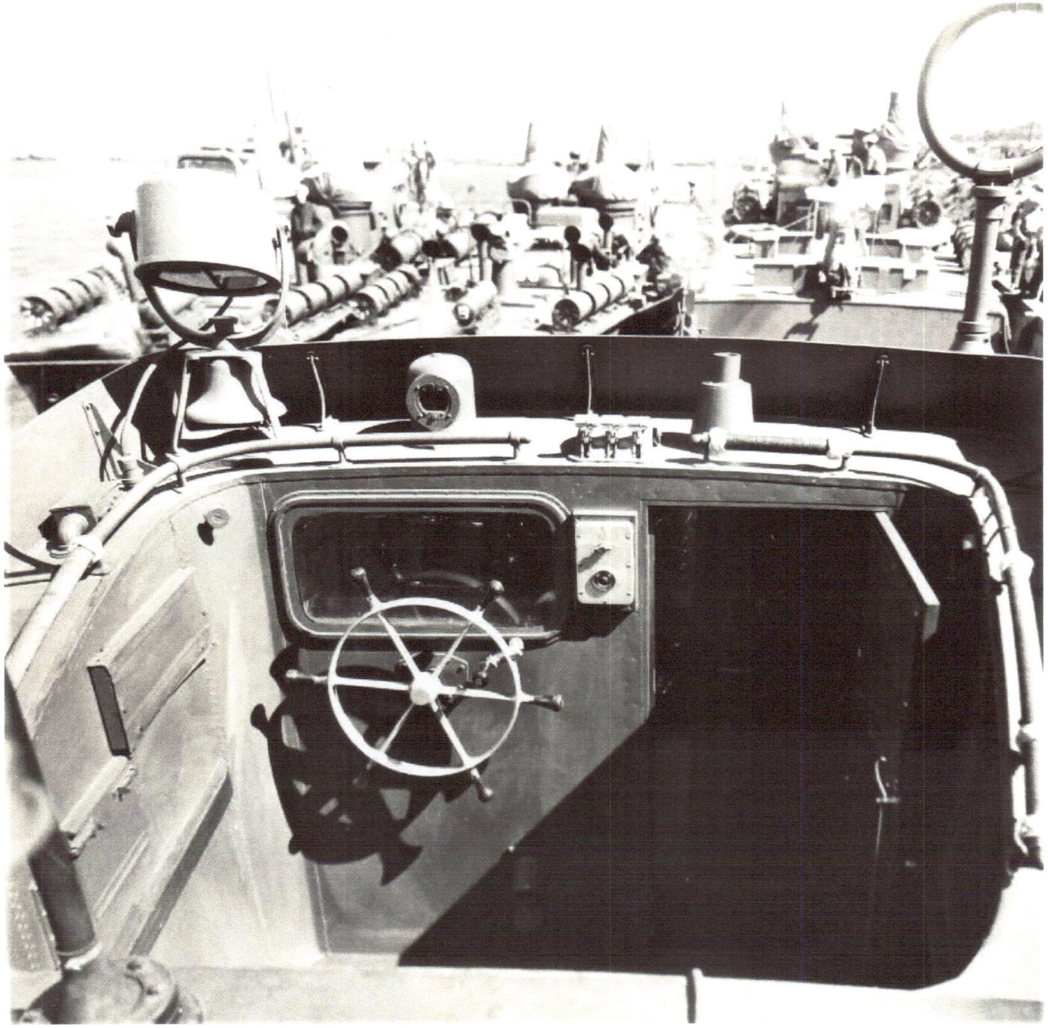

Figure 2-17. *A look inside the helm area of this second series boat, shows a bit larger structure than what was on the 70 foot boats. The wheel has a suicide handle for better control during high speed turns. Above the grab rail from left we see the search light, bell, binnacle, throttle controls, and the base for the torpedo director. Above the windshield we see the RDF, or Range Direction Finder. To the left on the cabin bulkhead, is the torpedo firing buttons, which are covered by canvas. The door to the right leads to the pilot house and forward crews quarters. (Charles B. Jones)*

Figure 2-18. *A look inside the port-side crews quarters. As one can see the area was not very large but very comfortable. In this photo we see the crews bunk area with the upper bunks being able to swing up when not in use. Water proof cushions complete the lower bunks. You get a good look at the crew locker area for stowing personnel gear, and the aluminum flooring plates, which could be lifted to clean the bilges. They were painted with a non slip material. Crews quarters was on both sides of the forward bow. (Irwin Chase)*

Figure 2-19. *Starboard side view of the PT-59 looking forward. Top bunks (also known as sette berths) as well as lower berths for the crew were here. The top bunks held by support chains could be unhinged and raised if needed. Through the open door was the boats water closet and sinks for the crew. Their was also a hatch that led to the forecastle of the boat. Storage of items was common in this area forward. (Gene Kirkland)*

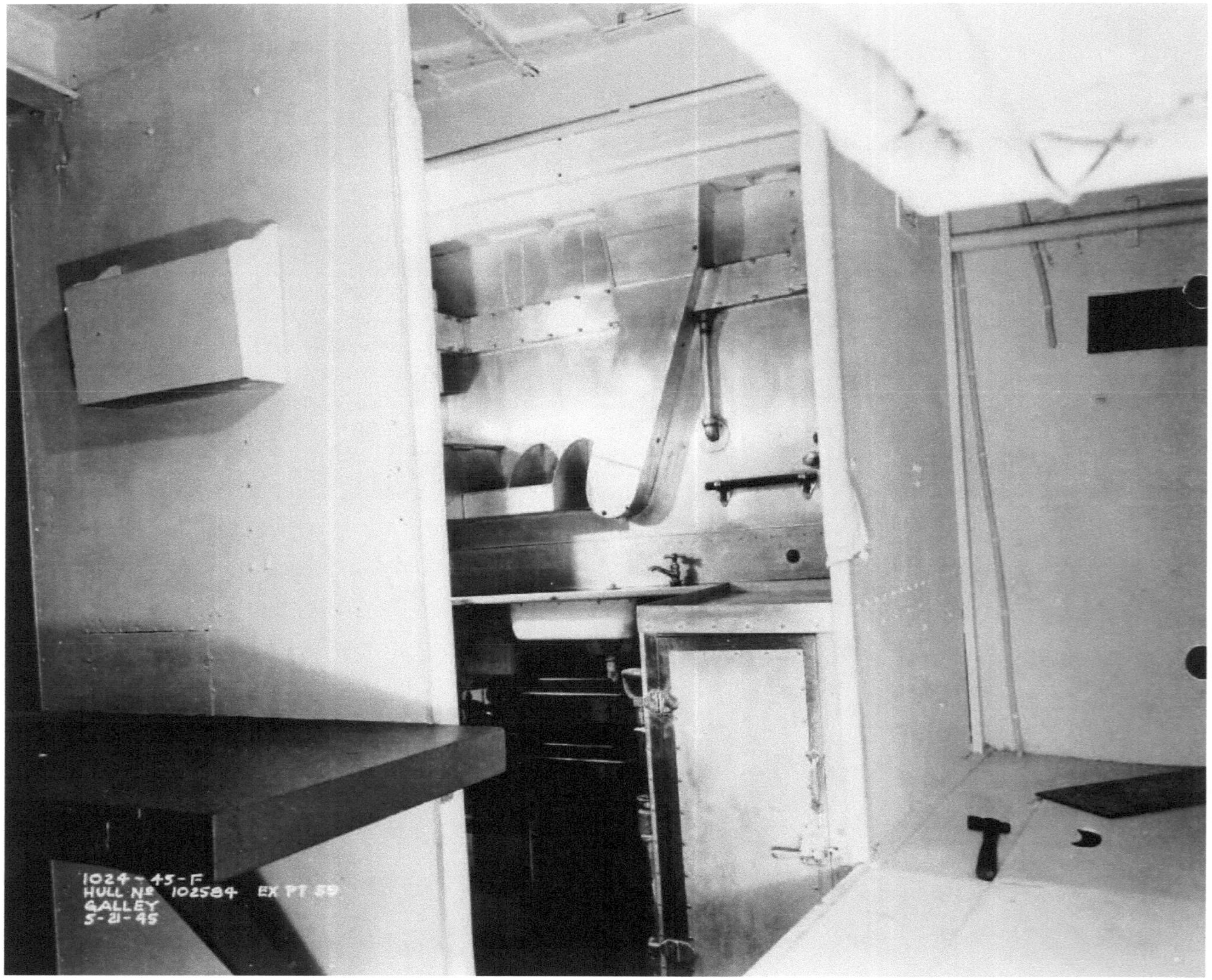

Figure 2-20. *A look at the small but functional Galley area onboard PT-59. The galley was located in the crews quarters separated by a dividing partition. The galley had a stove and sink area, and a place to keep dishes in a rack. They had a folding table for the crew to eat their meals. To the right of the sink is a small refrigerator that would keep some food stuffs cold. (Gene Kirkland)*

Figure 2-21. *This is the officers quarters on PT-59, and as one can see they were a pretty confining area. Two bunks made up the area for the Captain and Executive officer. The area had cabinet space for clothing and other necessary items. This photo shows the area looking forward on the port-side of the boat. Mattresses have been removed. (Gene Kirkland)*

Figure 2-22. *A look at the three 4M-2500 Packard Marine engines in the engine room, looking aft. These three engines would drive the boats three-bladed screws. The center engine was direct drive, with the two wings engines connected to vee drives. The engines were 1350 HP super-charged gasoline engines. The cooling water and lubricating oil in the engines were cooled by use of a heat ex changer in which the cooling medium was sea water. The engines gave the boats a top speed of 41 knots, but this was with a clean bottom. After being in action for several months, the speed would drop quite a bit due to bottom fouling, and general wear and tear on the engines. (Gene Kirkland)*

Figure 2-23. *The Dewandre Turrets that were installed on the 77 foot boats from PT-20-44, carried two 50 caliber aircraft machine guns. The power for operating the turret was compressed air at relatively low pressure, needing small pipes and fittings. The air transmits pressure using oil as the hydraulic medium. These mounts were completely automatic, as gunner's had only to manipulate the gun, to aim at the target when the turret and the yoke follow all the movements of the gun sufficiently, so that the aim can be kept on the target all the time. The draw back to these mounts was they would become useless if power were interrupted. They had a tendency to fog up while firing, causing limited visibility, and the tracking of this automatic weapon was too slow for the fighters they might encounter. These enclosed turrets were discontinued beginning with PT-45. (Glenville Tremaine)*

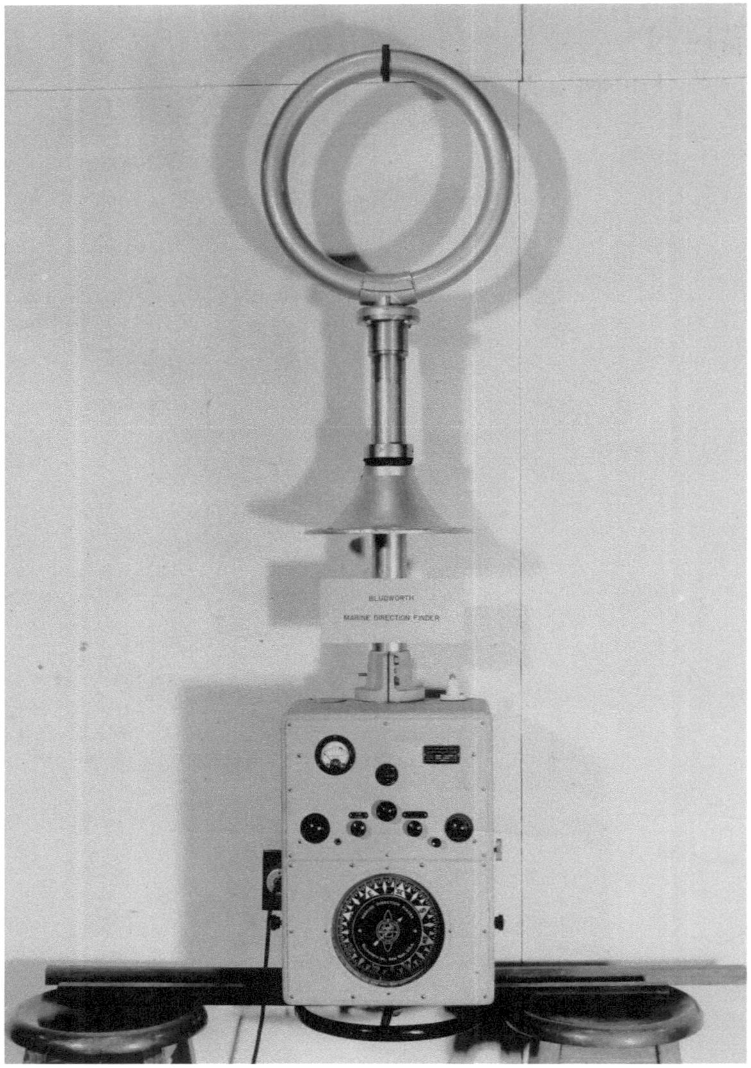

Figure 2-24. *A look at the Bludworth Marine Direction finder. This device was scheduled for installation on PT-68. The direction finder was basically used for determining the direction of the source of a radio signal. A direction finder (DF) can be used as a navigational aid. This is accomplished by measuring the direction (bearing) of at least two transmitters whose locations are already known.These devices were provided to Elco by the Simplex Bludworth Marine Instrument Company of New York. (PT Boats Inc.)*

Figure 2-25. *With the removal of the Dewandre mounts beginning with PT-45, we see the newer style replacement twin 50 caliber turret mount, in action on an unidentified Elco 77-footer. These mounts were manufactured by Colt, this particular mount was a simple manually operated Scarff-ring system. The Scarff ring was a type of machine gun mounting developed during the First World War. It allowed the gunner standing in an open gun tub to swivel and elevate his weapon quickly, and easily fire in any direction. The draw back was the small ammo boxes, which did not hold sufficient rounds that would be needed. (PT Boats, Inc.)*

Figure 2-26. *Nice view looking aft, shows the side by side gunner arrangement on the Elco 77 foot boat. This boat is still carrying the colt style twin 50 caliber machine gun mounts, using the scarff ring system. This would be changed to incorporate a better mount and increased sized ammo boxes, which would hold more ammunition. (Frank J. Andruss Sr.)*

Figure 2-27. *The boats could carry four 21 inch MK-VIII torpedoes. These torpedoes were fired electronically from the cockpit area, or by percussion at the rear of the torpedo tube, on top of the impulse chamber, should the circuit fail. Each tube had a metal dome that protected the torpedo from letting in salt water. These would be removed before firing. The torpedo was fired by using a black powder charge that when ignited would shoot the torpedo from their greased MK-18 tube. The torpedoes each had a 466 pound TNT warhead with a range of about 16,000 yards, at roughly 36 knots. These torpedoes were actually introduced in the first world war. (Irwin Chase)*

Figure 2-28. *This photo shows what happens, if the skipper pressed the button to fire the torpedo and nothing happens: At that point an Emergency percussion firing happens which is when personnel stationed at the tubes, strike the torpedo impulse chamber firing pin with a mallet, upon the order to fire torpedoes. It was standard practice to use this method of firing, whether electrical firing circuits were working or not. Here is one of the best photos showing the torpedo impulse chamber and the white button that would have been hit with a mallet to discharge the torpedo. Here you can clearly see in this photo the mallet held by this Sailor ready to strike that button for the impulse chamber. The humid climate of the Pacific would sometimes cause these charges to misfire, by not putting enough force behind the torpedo to fully eject it from the tube. Of course this method was only used on the boats when they had MK-18 tubes, and the MK-VIII torpedoes.(Frank J. Andruss Sr.)*

Figure 2-29. *A look at the standard 20mm Oerlikon cannon. This would be installed at the factory, beginning with the second series boats. Early installation included the MK-4 heavy destroyer pedestal mount. This weapon was drum fed and the canister held 60 rounds of high explosive shells and tracers. It had a range of 4,000 yards and was rated to fire at a rate of 450 rounds per minute. This cannon was used as an anti-aircraft weapon, which as heavier guns were introduced, would be moved to the bow of the boat. (Irwin Chase)*

Figure 2-30. *Here we see the early MK-II smoke generator which had two tanks. The smaller CO2 tank is upright at the head of the Titanium Tetra chloride tank and towards the center line. These smoke generators proved as much importance to the boats survival as did the weapons onboard. Laying smoke from the stern smoke generator which used Titanium tetra chloride which was a colorless, non-flammable, corrosive liquid. In contact with damp air it hydrolyzes readily, resulting in a dense white smoke. Hiding behind this smoke saved many a PT boat from being spotted by enemy ships. This early style smoke generator was used on some of the Elco 77 footers. (Irwin Chase)*

Figure 2-31. *The replacement for the early MK-II smoke generator was the Type C mount. Instead of two tanks, it was only one 35 gallon bottle mounted on the stern. It could be refilled, but could also be released and thrown into the ocean to confuse a chasing enemy destroyer. These generators could produce smoke from two minutes up to twenty-five minutes. (Frank J. Andruss Sr.)*

Figure 2-32. *A wonderful look at this Elco 77 footer as she has opened up her smoke generator. Notice the thick white smoke being made behind her, enough to hide the boat should she run into enemy ships. (Frank J. Andruss Sr.)*

Figure 2-33. *A look the depth charges available to the boats. The boats could carry four 21 inch torpedoes, and in some instances the aft torpedo tubes were removed, and depth charges were put into their place. They measured 27.625 inches long. Mark 6 and Mark 6 Mod. 1 each weighed 420 pounds complete-and are 17.625 inches in diameter. The deadly depth charges, thought to be used against subs, but more than once were used to drop in front of oncoming Japanese Destroyers, and used to sink stubborn enemy craft that would not sink with gun fire alone. Depth charges used to sink barges were seldom done, but it was tried. (Charles B. Jones)*

Figure 2-34. *A look at one of the changes to the turrets of the boats. These newer MK-17 mounts provided ammunition to the guns supplied from four boxes, two per gun, each having a capacity of 250 rounds, or 500 rounds per gun. This was a large upgrade from the former colt style mount. Canvas cover's were provided that fit over the open tops of two adjacently-placed ammunition boxes. The carriage of the mount provides unlimited train, which was accomplished by the operator exerting pressure, in the direction desired, on a winged back rest attached to the rotating carriage. The other noticeable change to the mount was around the guns, an aluminum combination spray shield- gun depression system, which kept the gunners from shooting into their own boats. (Charles B. Jones)*

Figure 2-35. *A nice look at this 77 footer which is docked at the Motor Torpedo Boat Squadrons Training Center. Notice the MK-17 machine gun mounts with the aluminum combination spray shield-gun depression system. You get a good look at the starboard side MK-18 torpedo tube, and impulse chamber at the back of the tube. The open sliding wooden hatch to the left of the photo leads down into the boats engine room. (Charles B. Jones)*

Figure 2-36. *Stern starboard side view of this seventy-seven footer, showing the 4 ventilators that will help bring in fresh air below decks. We also see the rear view of the forward MK-18 torpedo tube and the impulse chamber at the top. At the back of the torpedo tube is one of the boats ammo lockers. Notice this boat has been set up with eight MK-6 depth charges, four to a side. The boat carries her covered 20mm cannon, and we see just the frame work for the early style MK-II smoke generator. (Charles B. Jones)*

Figure 2-37. *Lt. Frank W. Rowe (second from right) was the Assistant Supervisor of Shipbuilding and it was he who would initiated, organized, established, and in October 1940 began operating a school at the Electric Boat Company, in Bayonne. This school was for familiarization and primary operational training of the officers and enlisted men, for the first established PT boat squadrons. More than 2500 officers and enlisted men would pass through this school. The school was located on 11 Schuyler Place in Bayonne. They are on the bow of one of the seventy-seven footers which has been moved to the outdoor boat shed. (Heather Rowe)*

Figure 2-38. *A rare look at the instant the MK-18 tube has fired it's MK-VIII torpedo. You can see the amount of power that has launched the torpedo, which can be seen just forward in all of the smoke. Gyro angles were usually set on the torpedoes before they were put into the tubes. These were closely checked to be sure that the settings were in the right direction to the standard setting for each particular tube. The tubes sat on training gears which would be cranked out so the torpedo when fired would clear the deck. (Charles B. Jones)*

Figure 2-39. *Wonderful stern shot of this 77 footer showing the powerful wake that was created by the design of the hull and the three powerful 4M-2500 Packard Marine engines. Engineers throughout the war tried to eliminate this wake, which was highly visible from the air, but had limited success. (Irwin Chase)*

Figure 2-40. *This second series boat PT-67, is slowly heading in to the Motor Torpedo Boat Squadrons Training Center, in Rhode Island. Notice she still carries her radio direction finder loop antenna on her forward cabin. The boat has two forward MK-18 torpedo tubes, and aft depth charges. The crew is dressed in underway weather protection gear as this photo was most likely taken in January 1942. (Bruce Marshall)*

Figure 2-41. *One of the most famous Elco 77 foot boats to come out of the war was PT-41. It would be commanded by Lt. John D. Bulkeley in the Philippines, with Squadron three. This is the boat that in March 1942, would take General MacArthur, his family and staff off the island of Corregidor, with Japanese forces beginning to surround the area. They traveled through stormy seas patrolled by Japanese warships and reached Mindanao two days later. From here MacArthur would take a plane to Australia. This wonderful model was constructed by Elco employee Donald Rosencrantz, who worked in the hull department. Don would build several models for the company and other dignitaries. Elco employees were thrilled when they learned that one of the boats they built, played such an important part in the war to date. (Donald Rosencrantz)*

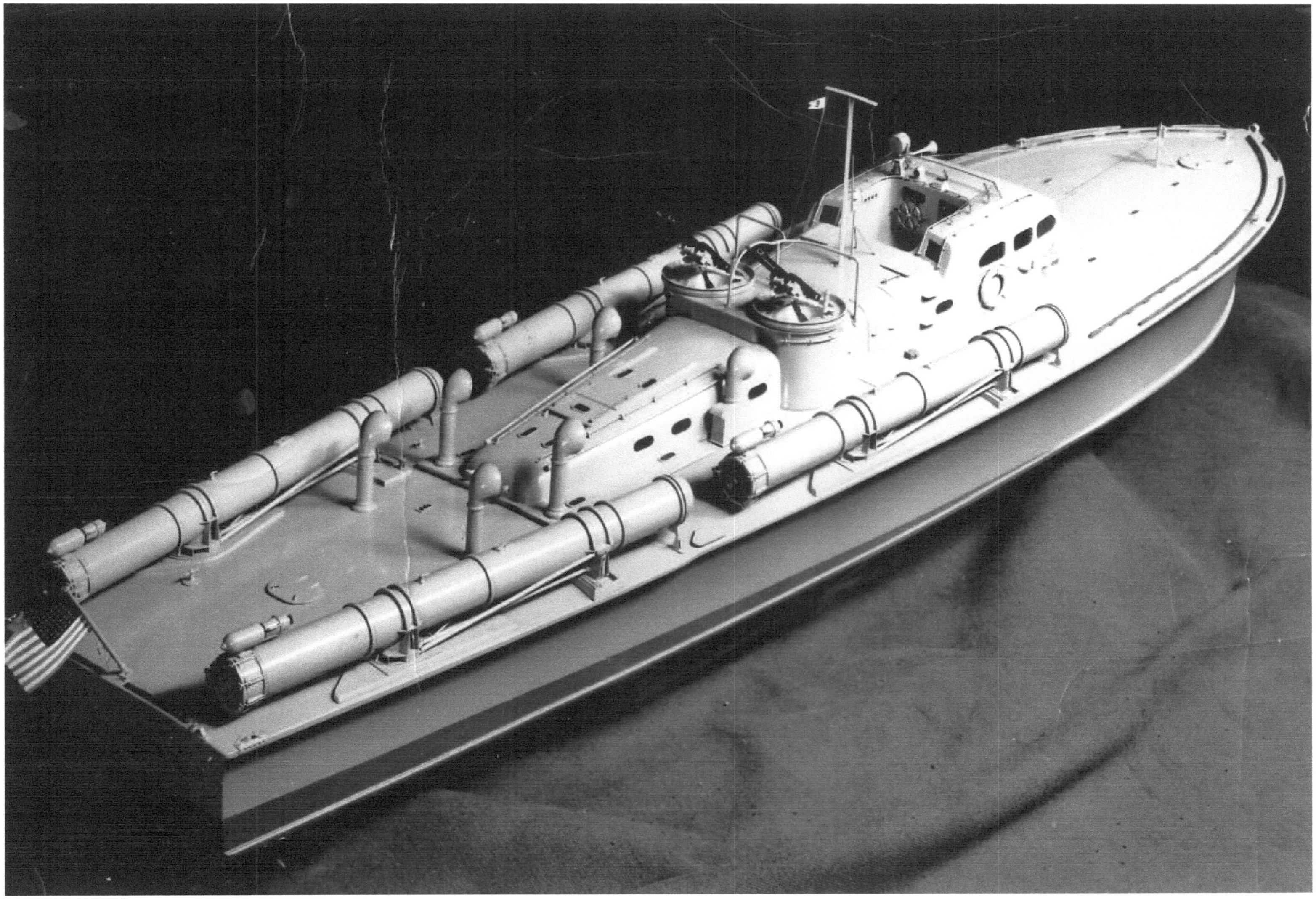

Figure 2-42. *A beautiful starboard side look at the PT-41 model. You can clearly see the amount of skill and craftsmanship that went into the building of a Rosencrantz built model. It should be noted that many of the parts on this boat were constructed of white metal, making his skill in building second to none. One of these boats was presented to Lt. John D. Bulkeley when he made a visit to the plant, after coming home from the Philippines. (Donald Rosencrantz)*

Figure 2-43. *Wonderful look at these boats as they skim past the New York Skyline. These first series boats are coming from the Kill Van Kull, likely from Elco. To the left out of the frame would be Bayonne, and in the background you can see the Brooklyn Bridge. The Brooklyn Navy Yard would be on the other side of the bridge. The boats would be making their way to be loaded sometime around December 12, 1941. The numbers on their cabins do not reflect the correct boat's that they are, and were set up this way because of spies. For instance PT-8 was actually PT-39, and PT-2 was PT-45. (Frank J. Andruss Sr)*

Figure 2-44. *Nice photo of PT-28 from Squadron 1, which is taking on gasoline. She is at the fueling dock at Pearl Harbor just prior to the Japanese attack. After the attack PT-28. along with two other boats returned to the Sand Island dock and sent rescue parties ashore. She would also participate in the Battle of Midway, and saw service in the Aleutian campaign. Later she was wrecked in a storm on January 12, 1943.(Frank J. Andruss Sr)*

Figure 2-45. *The cold months at the Motor Torpedo Boat Squadrons Training Center could be pretty brutal. The cold weather canceled many underway training classes, and became tough on the men. Here we see two of the second series 77 footers tied at the finger piers at the base. Both are set up in the same manner with two forward MK-18 torpedo tubes and eight aft depth charges. They both carry the 20mm cannon on the stern. The boat to the right is PT-65. Snow has covered the other boats numbers. (Charles B. Jones)*

Figure 2-46. *A nice look at several of the second series boats as they go through maneuvers in Long Island Sound, sometime in 1942. These boats would be stationed at the Motor Torpedo Boat Squadrons Training Center. They would also become famous as PT-62-68 were used in the film "Crash Dive" released in 1943. It stared Tyrone Power, Dana Andrews, and Anne Baxter. Elco employees were thrilled when they got to watch the movie in early 1944, when it played at the Dewitt Theater in Bayonne. (Ted Walther)*

Figure 2-47. *PT-64 is at the dock at the Motor Torpedo Boat Squadrons Training Center, while waiting for one of her MK-VIII torpedoes. The Hanson Torpedo Truck Crane was built by the Hanson clutch and machinery Company in Tiffin, Ohio. It was used in bringing the torpedo over from the warehouse where it will be eased into the MK-18 torpedo tube. (Charles B. Jones)*

Figure 2-48. *PT-39 has been lightered back to Elco from the Motor Torpedo Boat Squadrons Training Center. She sits in the boat basin on a barge and will be trucked to New York. The boat was introduced to the public at a large rally which was staged at noon on November 22, 1944. The boat was placed at the intersection of Wall and Broad Street's, and was used as part of the opening for the 6th War Loan Drive campaign, to launch the "Wall Street Fleet". She saw service with Squadron 3 at Guadalcanal before being transferred to the Motor Torpedo Boat Squadrons Training Center where she was used to train repair personnel. (PT Boats Inc.)*

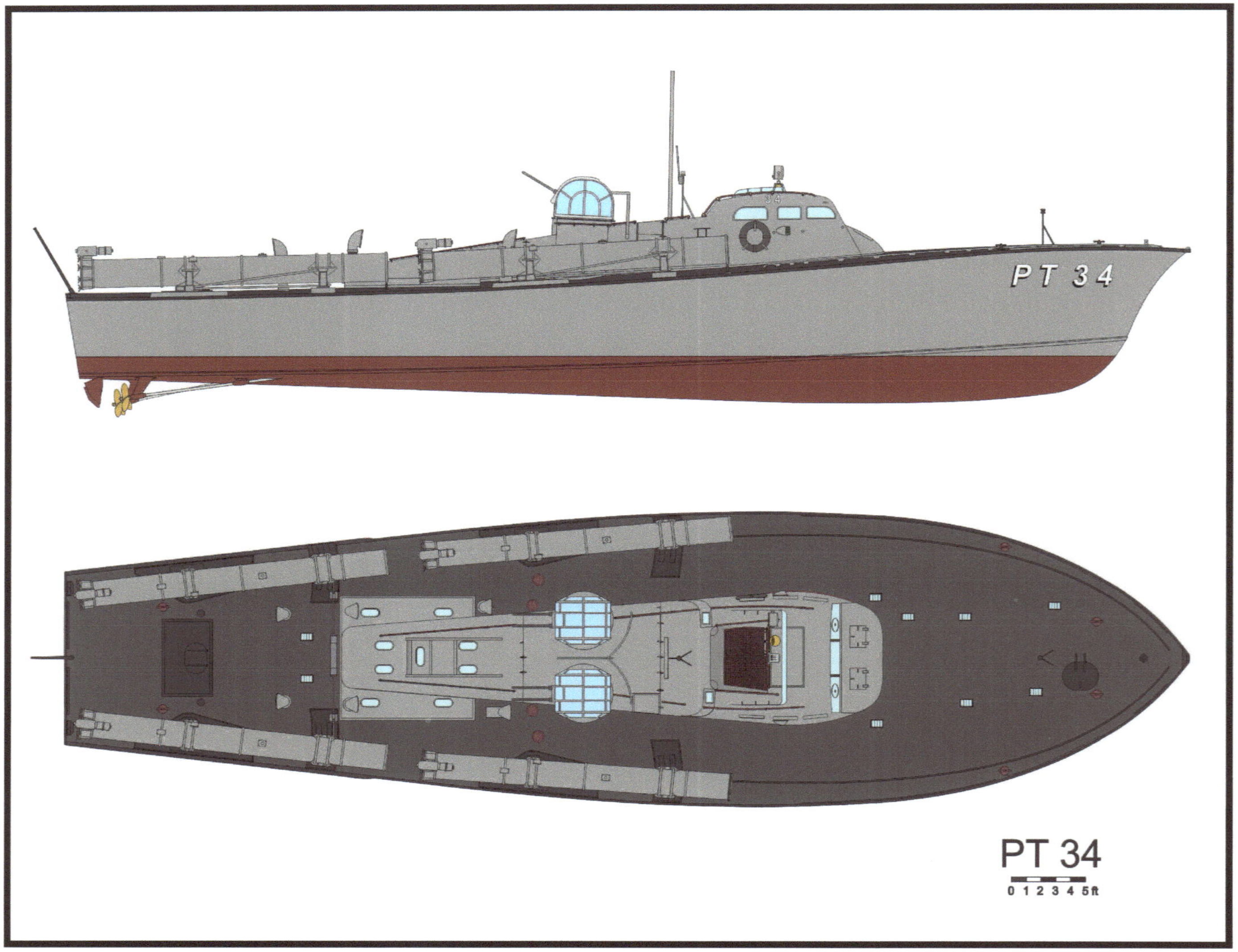

Figure 2-49. *Beautiful color drawing showing one of the first series boats. This is PT-34 as she looked coming off the assembly line. She carries her gray tones on her hull and deck and the Dewandre power turret, with four MK-18 torpedo tubes and four MK-VIII torpedoes. This boat would later serve with Squadron 3, under command of Lt. John Bulkeley. This was the only PT squadron engaged in the defense of the Philippines. Among its important missions was the evacuation of Gen. Douglas MacArthur from Corregidor in March 1942. (Drawing by Al Ross)*

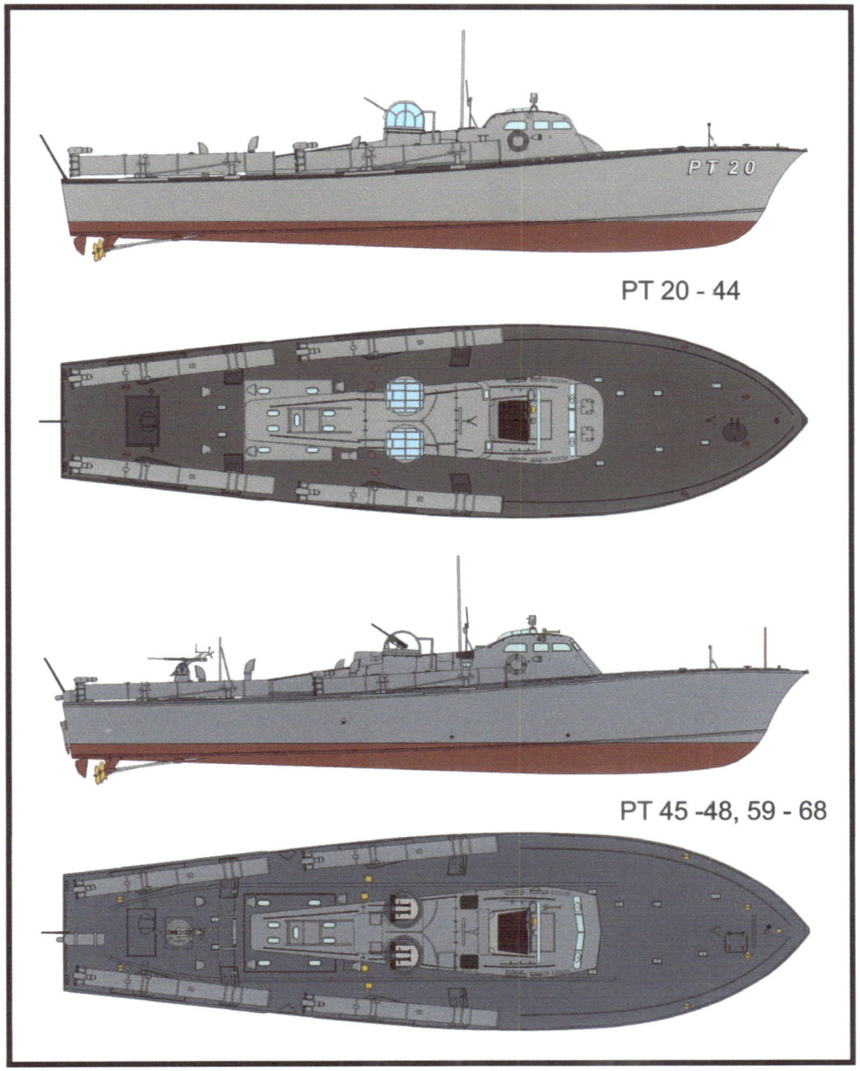

Figure 2-50. *This drawing showing the two different classes of the 77 foot boat. Gray tones were used for both classes coming off the assembly line, but one can see with the second series, that the Dewandre turrets have been removed and an addition of the 20mm cannon has been added to the stern. (Drawing by Al Ross.)*

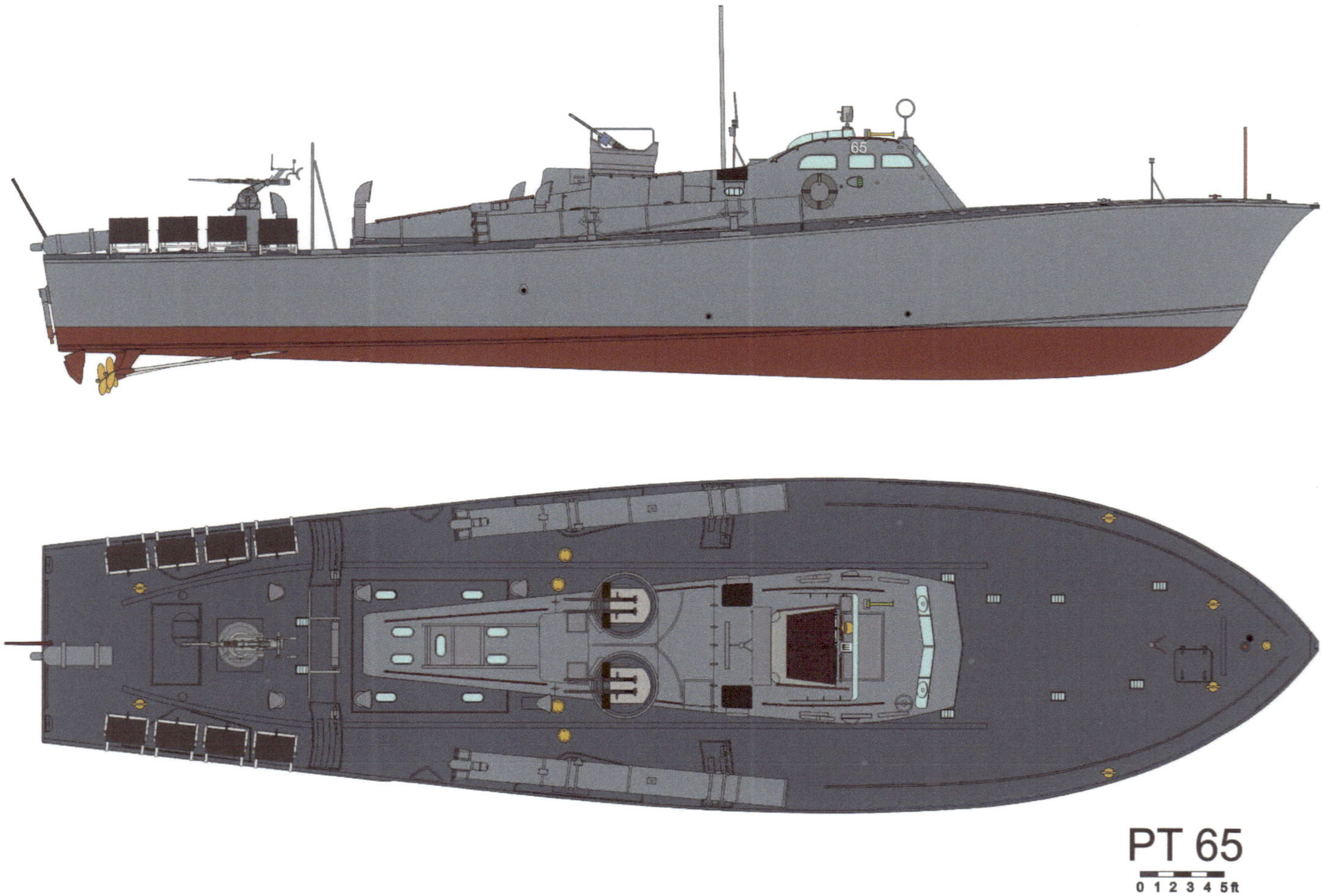

Figure 2-51. *Another fine drawing by Al Ross showing one of the second series boats. This one is PT-65, which has removed her aft MK-18 torpedo tubes and replaced them with the Type C Depth charges. One of the changes in the second series boats was the removal of the boot from the front of the pilot house, and a small platform aft of the cockpit, in which a whip antenna was mounted. They also strengthened the frame by adding deck stiffeners, which ran fore and aft on both sides of the boat. This boat would serve with Squadron 4, the PT Boat training school before being transferred to Squadron 5. (Drawing by Al Ross)*

Chapter 3

Elco Eighty-Foot

The new PT boat design for Elco would be the eighty-foot boat. The BuShips wanted a boat that was a maximum of eighty-two feet, which was felt could be transported easily and successfully.

The boat would still use the standard Packard Marine Engine power plant, but it was decided that they wanted a boat that could run silent at low power. Luckily for Elco, Irwin Chase their chief designer at the time, was an expert on mufflers. He had just designed three types of them for the British, which they felt was very satisfactory. It was also decided that these new boats would carry three thousand gallons of fuel.

The trial speed would be set at forty knots fully loaded, with the speed being sustained for at least one hour. The boat had to have accommodations for one officer and a crew of eight men, with spare accommodations for one officer, and forty-eight hours supply for provisions. They would set the cruising radius for five hundred miles.

The ordnance load did not vary greatly from the seventy-seven-footer, with specifications calling for four twenty-one-inch torpedo tubes, four .50 caliber machine guns, one thirty-two gallon smoke generator, and stowage capacity for one thousand round per barrel of .50 caliber ammunition.

The hull had to be of the lightest weight, consistent with adequate strength, stiffness, and durability for the service intended, with an eye to simplicity for mass production. Elco speed in production was increased by eliminating the molded day room, which had been a feature on the previous seventy-seven-foot design.

The eighty-foot boats were constructed similarly to the seventy-seven-foot boats in that transverse frames were prefabricated and placed upside down on a building jig. There were sixty-nine frames and the transom spaced differently depending on the load to be carried. The bottoms of the frames from the bow aft to the tank compartment are laminated of spruce, white oak, and mahogany. The remainder of the frames and the upper part of the forward frames are made of mahogany with a narrow white oak capping on the inner side.

After the frames are in place, longitudinal battens are let into the main frames from the bow to the forward end of the tank compartment where they are scarfed into a solid layer of mahogany planking which extends under the entire tank compartment. Battens are similarly scarfed into the after end of the tank-compartment inner planking, but from that point to the transom the battens run on top of the frames w th spruce spacers in between forming a solid flat foundation for the planking from the forward end of the tank compartment to the stern.

Secondary transverse frames and topside diagonal frames are similar to the seventy-seven-foot boats except that secondary frames are found in various parts of the boat where strain is excessive as well as in the forward three compartments. In the place of longitudinal deck battens, two longitudinal clamps are installed from bow to stern on either side of the deck openings for the superstructure. In addition, all the deck space between the superstructure openings has a longitudinal clamp running on the keel line. An eight-inch coaming is installed above the deck from the forward end of the superstructure to the after end of the engine room hatch. Diagonal wooden pads and stainless steel webbing under the torpedo tube foundations complete the deck framing.

Both deck and planking are bolted to a single continuous wooden member the entire length of the hull. The construction appears to be very strong and should give no difficulty. The deck is of double-planked mahogany, three-eighths-inch upper and five-sixteenths-inch lower, with marine glue and airplane fabric between the plank layers. The planks are laid fore and aft, each lower plank acting as a batten for the upper planks and fastened throughout, with brass screws.

The planking was double diagonal mahogany laid in the same manner as on the seventy-seven-foot boats. The total thickness of the topside planking is three-quarters inch, inner five-sixteenths inch, outer seven-sixteenths inch. The bottom planking totals one and five-sixteenths inch; inner, seven-sixteenths inch; outer, one-half inch. The fastenings to the main frames are Monel screws below the waterline and brass screws above and copper rivets or clinched nails are used for secondary fastenings. The deck houses were built of spruce frames with curved mahogany corners and covered on both sides with light plywood. The space between the frames in the superstructure was insulated with cork and the whole house is covered with airplane fabric, doped and painted. Hatches were provided to remove gas tanks and engines.

Potable water was carried in a single two-hundred-gallon tank located in the bilge under the chart room. The CO tanks were also installed in the same location with manual releases on the port side aft of the bridge and outside of the engine room hatch. No automatic discharge for the fire extinguisher system was provided.

The engineering installation was almost the same as the seventy-seven-foot boats. The heat exchanger, oil tank, and expansion tank for the center engine were in the port side of the engine room: those for the wing engines were in the conventional location. A seat for the engineer was provided on the starboard engine from which he can start the engines, observe the instrument panel, control the spark and scoops, and shift clutches. A square hatch with hold-down bolts could be lifted off to remove and change out engines.

Each Elco boat carried three aluminum gasoline tanks, consisting of two wing tanks, port and starboard, and the center tank. Each tank held a certain amount with the center tank carrying thirteen hundred gallons, and the port and starboard wing tanks carrying eight hundred fifty gallons each. Each boat could carry three thousand total gallons of high octane gasoline. These large tanks were carried under the day-room with the first boats not have self-sealing tanks, but this would change during the production run. A PT boat self-sealing gasoline tank refers to a design with a special self-sealing material that could automatically seal punctures caused by enemy fire, minimizing fuel leaks in combat situations; essentially, when hit by a bullet or shrapnel, the tank's inner lining would expand to plug the hole, preventing major fuel spills and potential fires.

The new 103 class boats would carry three 4M-2500 V-12 (super-charged) liquid-cooled, gasoline-fueled engines, made at the Packard Motor Car Company in Detroit. The engines continued to be a four-stroke, sixty-degree V-12 with a 6.40-inch bore and a 6.50-inch stroke, which brought it to 2,490 cubic inches, with four valves per cylinder. With an aluminum block with steel cylinder sleeves and weighing in at twenty-nine hundred pounds, the engine had a 6.4:1 compression ratio, and a centrifugal supercharger; later models were also fitted with an intercooler. The 103 class boat engines used one hundred octane fuel, rated at 1,350 hp, with later Elco boats having fifteen hundred hp as weights increased due to added weapons and crew.

One other item that was requested by the BuShips was the installation of depth charges and racks. Depth charges used on the new boats were of the type C-three hundred pound charge on a roll off rack. Depth charges used High explosives with a fuse set to detonate the charge, typically at a specific depth from the surface. Although hardly ever used against submarines, they were a deterrent against enemy ships, by dropping them at a very shallow depth in the path of the oncoming enemy.

Elco smoke generators were not mounted on the Elco 103 class until PT-115 to PT-126, thus becoming standard equipment coming out of Elco. They would be mounted parallel to the center line and was soon retrofitted to the earlier boats. The thirty-five gallon bottles mounted on the stern expelled titanium tetrachloride, which billowed into thick white fog when released into the atmosphere. It had a single threaded opening at one end, which was controlled by a valve. This valve is part of the bottle assembly, and the wrench that opens it is furnished with the original bottle (or generator) of smoke.

This wrench has a hinged handle so that a steel protective cap may be screwed onto the bottle over the valve assembly with the wrench attached when the generator is not in use. To spray the smoke liquid from the generator, a smoke pipe was furnished. This pipe was about two feet long and is bent slightly. On the lower end it carries a fitting to connect the pipe to the nipple on the generator valve assembly. A hinged wrench for this fitting is clipped to the smoke pipe. At the upper end the pipe is threaded to receive a single nozzle.

Four nozzles were provided, and were quickly interchangeable. They are marked for three min., five min., eight min., or twelve min.; the time the smoke will last in a continuous screen, using the respective nozzle. The eight-min. nozzle is for use under average condition and the twelve-min. nozzle is for night use. If the weather is very windy either a three or five-min. nozzle would produce a better screen. With a few turns of the wrench the tanks could be discarded overboard and with the tanks ability to float would act as a decoy for attacking ships.

Once again the gun turrets would be the aircraft M-2 .50 caliber twin machine guns, with changes from the earlier boats, being two turrets, one port side, attached to the day room cabin and the other starboard, opposite the bridge. They used the Mark 17 mount, which included the Cradle, Mark 9 and Carriage, Mark 9. The carriage of the mount provided unlimited train, which was accomplished by the operator exerting pressure, in the direction desired, on a winged back rest attached to the rotating carriage.

Ammunition to the guns was supplied from four boxes, two per gun, each having a capacity of 250 rounds, or five hundred rounds per gun. Ammunition boxes were carried on supports attached to the rotating carriage structure. Ammunition was always packed in the boxes, so that the double loop of the link will be leading and, therefore, will be fed into the gun first. This was important because the feed mechanism of the gun will operate with a greater belt load if the double loop is fed first, for then both the cartridge-feed pawl and the holding pawl will always bear against a single thickness of the link thereby properly spacing the belt. These guns were used for anti-personnel and light anti-ship fire.

Elco boats had six mufflers attached to their transom, three on the port side and three on the starboard side. They were constructed of mostly copper and bronze. Mufflers were used by PTs for silent running by feeding the exhaust out under the water at low speeds. They each had butterfly flap valves fitted in the main exhaust pipes after the muffler take off which were operated by a lever system to allow straight through exhaust for running at higher speed or to direct the exhaust down through the mufflers to silence the boat at low speed. When exhaust was sent out beneath the water the boats could run in basically a silent mode, making them less noticeable to the enemy. They were constructed by the Van Blerck & Son company of New York. This photo shows the boats port-side mufflers.

The PT boat was designed to carry four torpedoes in tubes and to discharge them by expansion of ignited gun powder. There were two tubes on each side. Tubes train inboard for cruising and loading, and train outboard for expected action and firing. The forward tubes train outward to an angle of 8 degrees 30 minutes with the ship's centerline, the after tubes to an angle of 12 degrees 30 minutes. All torpedoes are fired forward and at the stated angles with the centerline. Before loading the torpedoes into the tubes, it was intended that the gyro in each torpedo be set to the respective angle of the tube into which it is to be loaded. The direction of setting must be such that after leaving the boat the gyro will correct the course of the torpedo to that of the boat at the instant of firing, plus the spread angle. The torpedo used would be the tube mounted MK-VIII, a twenty-one-inch WW1 vintage model, with a top speed of only twenty-seven knots.

It still carried the dismal record of being unreliable (60 percent) dud rate, with a low speed. Another issue was that the tubes were heavily lubricated with oil and grease to ensure that the torpedo was smoothly launched from the tube. Sometimes during launch, the charge would ignite the lubricants in the tube causing a fire emitting black smoke that would reveal the location of the boat. The gyro in the torpedo was sensitive when it was first launched and it would need to be launched from an even keel, otherwise the torpedo would lose stability when it hit the water. The other issue concerned the launching mechanism.

The tube was made of steel, fabricated by welding. It pivoted on an aluminum alloy turntable and turntable base. The forward end is mounted on an aluminum alloy support which travels on corrosion-resistant steel plates and is guided by brass Z-bars, cadmium plated. The turntable and the forward support are joined with steel tubing rails. The tube does not recoil. When the charge is fired, pressure is built up in the chamber and gas is fed into the tube through a throttling orifice located between the chamber and the tube.

The tube pressure reaches a maximum soon after torpedo movement begins and decreases gradually to zero as the torpedo clears the muzzle. The firing mechanism was designed for either percussion or electric firing, depending on the type of primer used. The firing pin, which is insulated, is wired to an electric circuit. When the striker bolt is in firing position, it forces the firing pin against the primer, completing the circuit from the control station. The primer may then be fired by hitting the striker knob or by closing the electric circuit with the switch provided at the control station.

Later these torpedo tubes would be removed from the design, with the advent of the newer MK-XIII model torpedo, which would be carried in lightweight MK-1 launching racks, at a total saving of more than fourteen hundred pounds each. The shorter Mark XIII also took up less deck space, which would be launched using racks, thus eliminating the heavy steel tubes. The Mark 13 also had the advantage of a significantly larger warhead, six hundred pounds vs. 466 pounds containing the significantly more powerful explosive Torpex, which was approximately 1.5 times more powerful per unit of weight than TNT. The Mark 13's shorter range and slower speed were considered acceptable trade-offs for boats that usually operated at night and relied on stealth to reach firing position. This system began coming out of the Elco factory in late 1943, with the earlier boats also being changed over to the lightweight rack and torpedo.

The 103 class boats coming out of the Elco factory would carry the twenty-mm Oerlikon antiaircraft cannon on the stern. The weapon needed no external power or cooling supply and could be bolted down in any available deck space. The gun could also be operated by almost anybody with only a limited amount of training.

The air-cooled twenty-mm Oerlikon had some unusual features not found in other automatic weapons, including an open breechblock and a spring-recoil/counter-recoil system. It fired a 0.3-pound explosive shell supplied from a detachable magazine that held sixty rounds; expending all of them took just seven and a half seconds. The mount on which the gun rested was the MK-4 mount. It was equipped with open ring-sights for aiming. Pointing and training were controlled through a set of handlebars and shoulder rests. The gunner had to lead the target in both traverse and elevation to achieve hits. To aid in spotting, a tracer was usually included in every fifth round. Later this gun would be removed from the stern in favor of the forty-mm gun, a harder hitting weapon. The addition of the Bofors forty-mm cannon, significantly enhanced the defensive capabilities of PT boats, especially against aerial attacks.

The galley on the Elco eighty-footer was spacious in regards to other boats. It was located to starboard from the center line between watertight bulkheads and entered forward through a door. The small sink had over-all dimensions of fourteen by twenty by six inches and was a commercial porcelain enamel type, with strainer and drain plug.

The galley had cupboards made of western red cedar with door frames of mahogany, and two shelves each with ledges, were built below the sink board and counter. Cupboard space outboard above refrigerator were fitted with partitions and doors, also racks and cutlery drawers to stow specified galley equipment. The small stove was a twenty-four-volt D. C. unit with a latch for anchoring the oven in place, and provided with rough water utensil keepers. A small refrigerator was also provided that was roughly eight cubic feet, and had its own evaporator unit and ice trays.

The boat contained two toilet's or heads. There was the officers head, with the crews head being in the fore-peak of the boat. They had a stainless steel sink with which a pump was used to bring water to the basin. This compartment was fitted with medicine cabinet on the after bulkhead, and a frame-less mirror on the forward bulkhead, also, with three grab rails, located on the forward and after bulkhead. Toilet fixtures included a paper holder and soap dish, the grab rails being placed as to function as towel bars. The toilet was a hand pump that would release into a holding tank that released into the sea.

The forward crews quarters provided comfortable bunk space and spacious quarters. They contained upper bunk areas both on port and starboard sides of the forward hull, with a mahogany drop down table which would be used for crew meals. The crew were afforded mahogany painted lockers with air holes, to keep mildew from forming on clothing. The area was painted with a bright white paint, that made it easier to see below decks. There were also over-head lights in this compartment for night use. On the eighty-foot design, this was the most spacious compartment on the boat, with thick waterproof cushions.

The day room cabin onboard the 103 class boats was considered at the time, to be that of a lounge area for the crew. It was pretty spacious and had six cabin windows, three to a side, for more air. It was located above the fuel tank compartment between watertight bulkheads, with a non watertight door to forward quarters, and sound-insulating door and hatch in after bulkhead to engine room. All of the furnishings above the day room floor were easily removable, so that flooring and tank side panels could be removed for inspection or service of the tanks. In the starboard forward corner, a hatch of at least twenty-two inches square was cut, with a hinge back top, to provide an exit from the day room, using a steel ladder with tubular rungs mounted on the house side. It had two couches that faced one another for comfort in mind. Potable water was carried in a single two-hundred-gallon tank located in the bilge under the chart room.

There was an area known as the Lazarette. It was the last compartment at the stern of the boat, and was originally designed for a work space. It was accessible through a hinged deck hatch with metal ladder to floor, and through a watertight door in after bulkhead of engine room. In this space was the auxiliary steering apparatus, where if need be they could rig an emergency tiller. It would become a perfect spot for storing gear, such as oil and paint cans, lubrication gear, smoke generator parts, and even extra ammunition.

The 103 class boats produced by Elco would run from 103–196. At that time they would produce a second class of eighty-footer, from PT's 314–367. A third class of boat would be constructed from PT's 373–383. Several changes with each class of boats were implemented, with a fourth class being PT's 486–563. A fifth class of boat would be PT's 565–624. The 624 being the last boat to be excepted by the US Navy. Elco would continue making boats with the last class being PT's 731–790, which were knock down kits produced and shipped to Russia. PT's 761–790 were canceled and never shipped.

"SO Radar" refers to the type of radar system that was later installed on many US Navy PT boats, giving them a significant advantage in night operations by allowing them to detect enemy ships at a distance, particularly for targeting supply barges and other vessels in the dark; "SO" stands for "Surface Search" indicating its primary function of detecting ships on the water's surface. The radar allowed PT boats to locate and attack enemy ships at night without relying solely on visual sightings, giving them a tactical advantage. The radar was primarily produced by the Raytheon Company, and had a relatively limited range, typically around seventeen nautical miles.

There is no doubt that the Elco eighty-foot design, was the most advanced that was produced by the engineers, and the workers of the plant. Elco was the dominant builder of PT boats, producing more than any other small boat builder, with 399 by wars end. Later boats mounted one Bofors forty-mm gun, and a thirty-seven-mm auto cannon coming from the factory. Engineers were constantly improving the quality of the boats and the men and women can certainly be proud of the boats that they built for the war effort.

Figure 3-1. *Eighty-foot designed PT boats take up plenty of space inside Building 21 under construction. Here PT-103, the first of the eighty-footers begins life from this jig. The hull will take shape as frames are added. (Irwin Chase)*

Figure 3-2. *Frames are being added to the jig on this boat, which consist of 69 frames and the transom spaced differently depending on the load to be carried. The bottoms of the frames from the bow aft to the tank compartment are laminated of spruce, white oak, and mahogany. The remainder of the frames and the upper part of the forward frames are made of mahogany with a narrow white oak capping on the inner side. The boat to the left has the mahogany planking installed on her hull. Boats forward are in another phase of assembly, and have been turned over to begin work on the deck. Notice the two workers handling one of the numbered frames that shows a water tight door, which will lead to other compartments. (PT Boats Inc.)*

Figure 3-3. *The boat to the lower right of the photo has the frames are in place, longitudinal battens are let into the main frames from the bow to the forward end of the tank compartment where they are scarfed into a solid layer of mahogany planking which extends under the entire tank compartment. Battens are similarly being scarfed into the after end of the tank-compartment inner planking, but from that point to the transom the battens run on top of the frames with spruce spacers in between forming a solid flat foundation for the planking from the forward end of the tank compartment to the stern (Irwin Chase)*

Figure 3-4. *The first layer of planking is attached to the hull. Planking was Mahogany in two layers, laid diagonally in opposite directions. The hull when completed will have taken some 400,000 rivets and screws to complete the job. (Irwin Chase)*

Figure 3-5. *A popular custom was the turning of the hull, after the planking process was completed. Workers from different Departments were chosen for this ceremony as well as other dignitaries, war bond sponsors, and Navy personnel. With the hull turned over using heavy winches it could now proceed to the next phase of assembly. Here Electrical Department worker Gladys O'Reily was chosen to turn this hull, which has been adorned with a sign, thanking the people of the Virgin Islands for playing their part in the war effort. (Irwin Chase)*

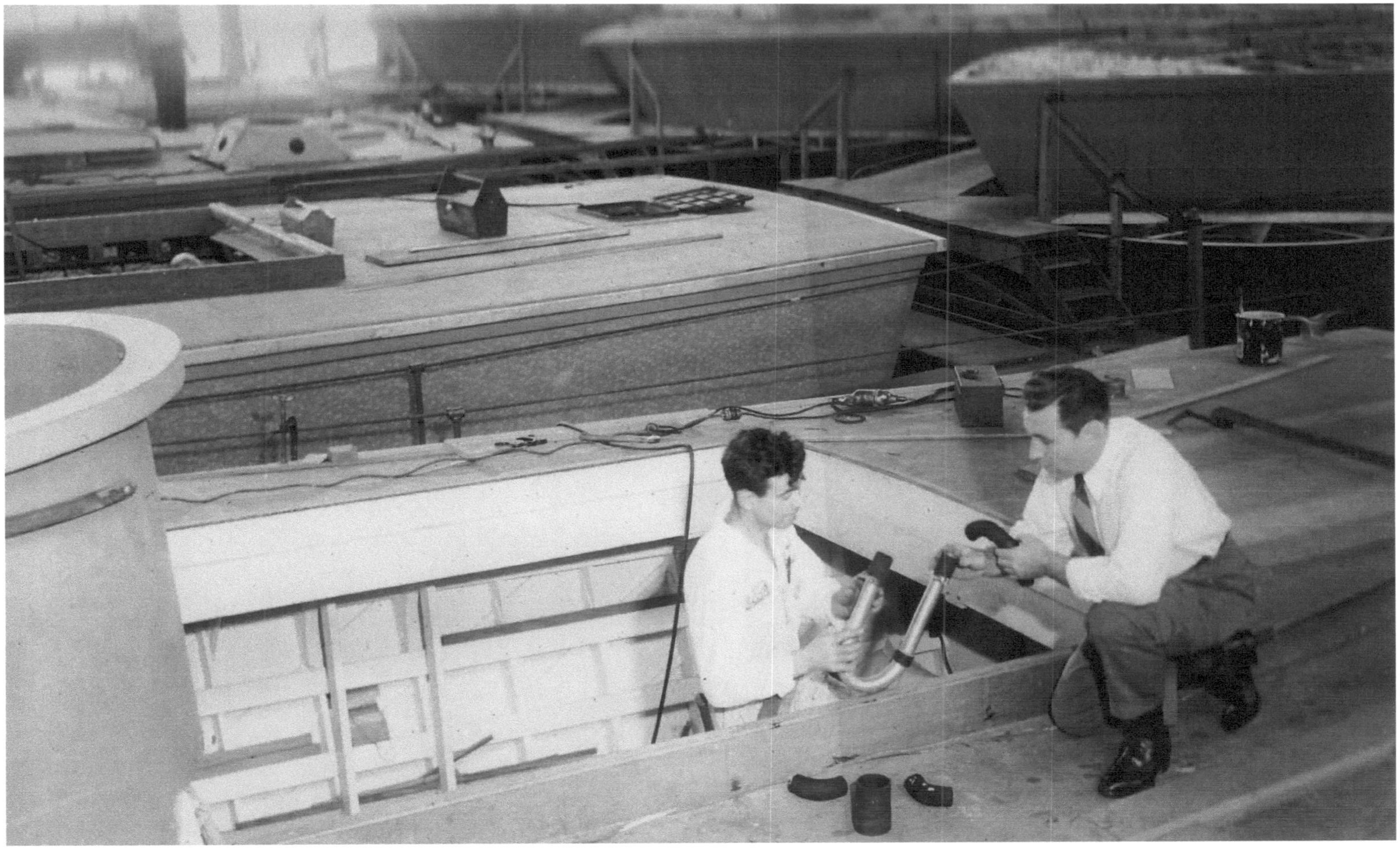

Figure 3-6. *Worker in the assembly area is going over the plumbing, which will be installed in the engine room of this boat. All Piping and fittings were of aluminum to all fixtures. One of Elco's mechanical engineer's is checking out the piece making sure connections will hold tight under combat conditions. (PT Boats Inc.)*

Figure 3-7. *With the hulls now turned over, more phases of assembly can be completed. Here we see the helm area which has been added to the boat, while the boat next to it, workers are assembling the boats day room cabin. (Irwin Chase.)*

Figure 3-8. *Elco worker completing the boats day room cabin, using an electric hand drill. He is sinking the hundreds of screws in this assembly process. The cabin was built of spruce frames with curved mahogany corners and covered on both sides with light plywood. The space between the frames in the superstructure is insulated with cork and the whole house will be covered with airplane fabric, doped and painted. (PT Boats Inc.)*

Figure 3-9. *In Building 21 we get a good look at one of the steel cradles on wheels. These cradles were fabricated from steel I-beams to move the boats around the yard. The hull supports were shaped wood beams laying on top of steel I-beams. The structure sits on wheel sets, with "tillers" at each end, making the wheel sets swivel. Once the boats are turned over, they are placed into the cradle and will remain here right up until being rolled out of assemble and into the launching crane, where it will be gently placed into the boat basin. These cradles weighed just over 11 tons. (PT Boats Inc.)*

Figure 3-10. *One of several shops that were set up in the assembly building. This one showing several work benches in which intricate parts of the boat are being worked on. To the right of the photo we see five 4M-2500 Packard Marine Engines that gave the boats life. The boats in the background assembly area tower over the shop area. The boat third from the left is PT-322. (PT Boats Inc.)*

Figure 3-11. *This women from the paint shop is getting the Eclipse paint sprayer ready for spraying the enteral bulkheads. The entire interior was painted with one coat of primer, glazed as required, and two coats of fire retardant white paint. An extra coat of enamel was applied to the surfaces adjacent to passageways, areas of traffic, and confined spaces as directed by the Supervisor, where personnel traffic would render the fire retardant paint difficult to maintain. The sprayers used at Elco were provided by the Eclipse Air Brush Company in Newark, New Jersey.(PT Boats Inc.)*

Figure 3-12. *PT boats ran three 4M-2500 Packard Marine Engines and in so doing the boats had three props each. The boats at delivery had one 28" x 28" propeller for the center shaft, and one 28" x 29" propeller for each wing shaft. Each boat would be given one center prop and two wing props as spares when delivered with each boat. Each prop was unpainted manganese bronze with right hand turning props. With right-hand turning propellers there was a slight torque, or tendency for the vessel to go to the left with her engines ahead and rudder amidships. (PT Boats Inc.)*

Figure 3-13. *Elco boats had six mufflers attached to their transom, three on the port side and three on the starboard side. They were constructed of mostly copper and bronze. Mufflers were used by PT's for silent running by feeding the exhaust out under the water at low speeds. They each had butterfly flap valves fitted in the main exhaust pipes after the muffler take off which were operated by a lever system to allow straight through exhaust for running at higher speed or to direct the exhaust down through the mufflers to silence the boat at low speed. When exhaust was sent out beneath the water the boats could run in basically a silent mode, making them less noticeable to the enemy. They were constructed by the Van Blerck & Son company of New York. This photo shows the boats port-side mufflers. (PT Boats Inc.)*

Figure 3-14. *Auxiliary Firemen have taken up station in the boats forward crew's quarters. Here they are vacuuming out the bilges, making sure all saw dust and other materials have been removed before delivery. The white doors with the holes in them, are for the crews belongings, the holes to keep clothing from collecting mold and mildew. (Irwin Chase.)*

Figure 3-15. *Each Elco boat carried three aluminum gasoline tanks. From left we see one of the wing tanks, the center tank, and the second wing tank. They have just been painted before being lowered into the boat. Each tank held a certain amount with the center tank carrying 1300 gallons, and the port and starboard wing tanks carrying 850 gallons each. Each boat could carry 3,000 total gallons of high octane gasoline. These large tanks were carried under the day-room with the first boats not have self sealing tanks, but this would change during the production run. A PT boat self-sealing gasoline tank" refers to a design with a special self-sealing material that could automatically seal punctures caused by enemy fire, minimizing fuel leaks in combat situations; essentially, when hit by a bullet or shrapnel, the tank's inner lining would expand to plug the hole, preventing major fuel spills and potential fires. (Irwin Chase)*

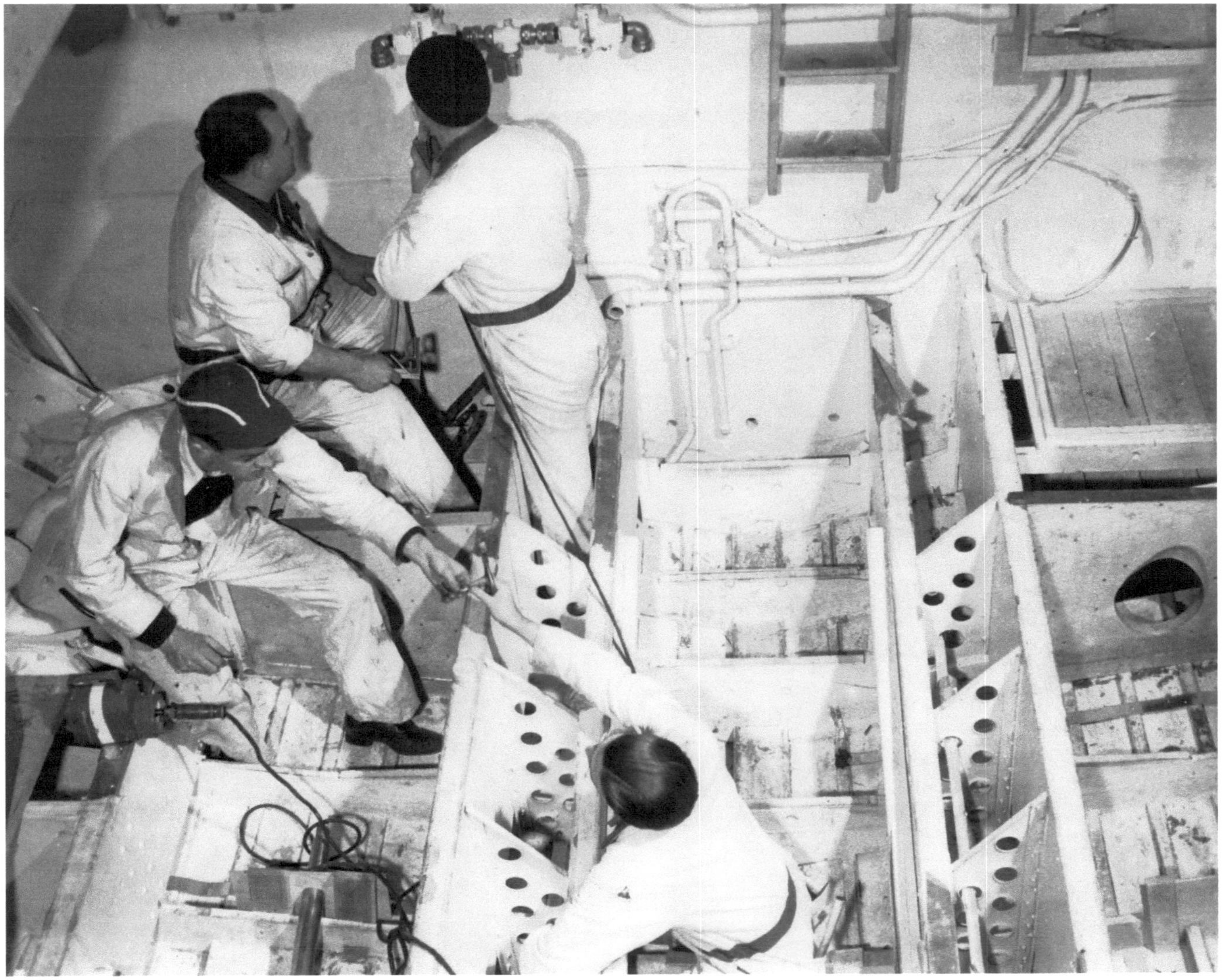

Figure 3-16. *Nicknamed the Elco Engine room gang, these workers are busy inside the boats engine room compartment. The two workers in the rear are working on the plumbing which is attached to the aft watertight compartment. The worker seated with the hat on has a drill next to him, while the other worker is handing him a combination square. Notice the girders and engine beds that will soon hold the three 4M-2500 Packard Marine Engines. (Ed Behney)*

Figure 3-17. *Engines were delivered from the Packard Motor Car Company in Detroit, to Elco. They were packed in this type of crate with instructions to remove the lag screws from each side and door top. The engines came secured to a skid and would carefully be removed using a winch. Engines would come by rail which ran to the back of the Elco property. (PT Boats Inc.)*

Figure 3-18. *Each boat carried three 4M-2500 Super-charged gasoline engines. These power plants were a four-stroke, 60-degree V-12 engine. They had a 6.40 inch bore and a 6.50 inch stroke, which brought it to 2,490 cubic inches, with four valves per cylinder. They had an aluminum block with steel cylinder sleeves, weighing in at 2,950 pounds. Each engine carried a centrifugal super-charger, with a Holly 1685F Aircraft carburetor supplying the fuel. The engines were dry sump with an external oil tank which had baffles to reduce foaming. The engines were fresh water cooled via fairly small tube type heat ex-changers, as were the V-drives. The first engines developed 1,250 hp, but as the war progressed, we would see this increased to 1,500 hp. (Frank J. Andruss Sr.)*

Figure 3-19. *The early eighty-foot boats had these MK-18 Mod. 1 torpedo tubes. The Mark VIII or (Mark 8) with the smaller warhead and the Mark XIV or (Mark 14) with the larger warhead were both used on USN PT boats. Both with 21 inch diameter to fit the standard USN tubes. These MK-18 Mod. 1 tubes were built by the Whitlock Manufacturing Company in Hartford, Connecticut. Launching the torpedoes from these tubes was done by utilizing a black powder launching charge. Sadly these torpedoes were dismal in combat, as they were an obsolete relic from World War One. They had a slow speed (27 knots), a small warhead containing 300 pounds of TNT, a tendency to run wild when set for shallow depth, and a defective exploder that would detonate prematurely, if it exploded at all. It was later estimated that sixty-three percent of the MK-VIII torpedoes fired from these tubes had service issues. It wasn't until mid 1943 to early 1944, that the boats torpedo's would be upgraded and these tubes removed from the boats. (Frank J. Andruss Sr.)*

Figure 3-20. *Here we see one of the heads that was on the 80 footer. This was the officers head, with the crews head being in the fore-peak of the boat. Notice the stainless steel sink with which a pump was used to bring water to the basin. This compartment was fitted with medicine cabinet on the after bulkhead, and a frame-less mirror on the forward bulkhead, also, with three grab rails, located on the forward and after bulkhead. Toilet fixtures included a paper holder and soap dish, the grab rails being placed as to function as towel bars. Access to the compartment was through doorway with curtain. The toilet was a hand pump that would release into a holding tank that released into the sea. (Frank J. Andruss Sr.)*

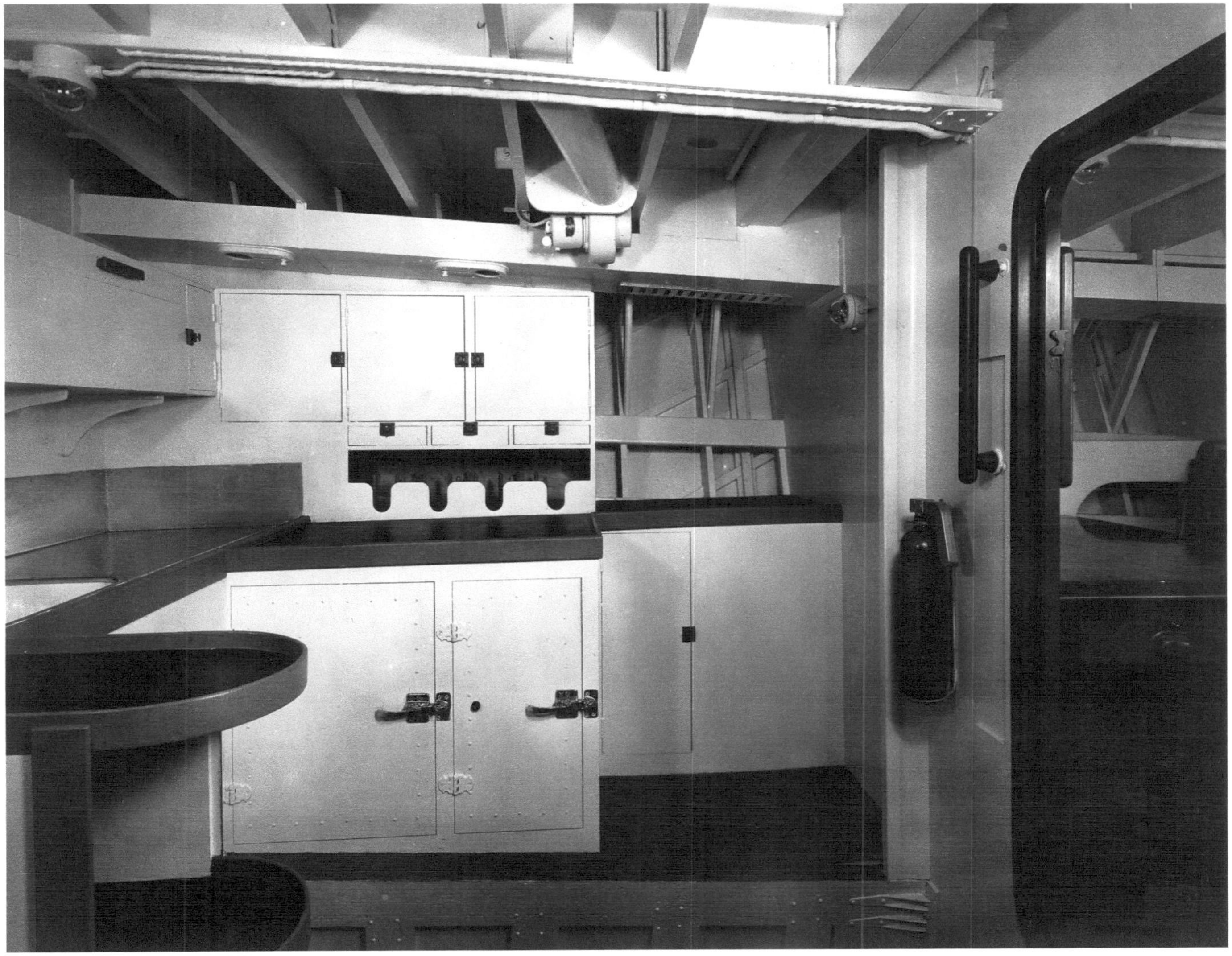

Figure 3-21. *The galley on the Elco 80 footer was spacious in regards to other boats. It was located to starboard from the center line between watertight bulkheads and entered forward through a door. The small sink had over-all dimensions of 14 by 20 by 6 inches and was a commercial porcelain enamel type, with strainer and drain plug. The galley had cupboards made of western red cedar with door frames of mahogany, and two shelves each with ledges, were built below the sink board and counter. Cupboard space outboard above refrigerator were fitted with partitions and doors, also racks and cutlery drawers to stow specified galley equipment. The small stove was a 24-volt D. C. unit with a latch for anchoring the oven in place, and provided with rough water utensil keepers. A small refrigerator was also provided that was roughly 8 cubic feet, and had its own evaporator unit and ice trays. (PT Boats Inc.)*

Figure 3-22. *Two sailors look to be enjoying the comfort afforded them by comfortable bunk space and spacious quarters. Above them are the upper bunk areas, and to the left of the photo we see the mahogany drop down table which would be used for crew meals. Upholstery, mattresses, and cushions for below deck comfort were provided by the Atlantic Pacific Mfg. Corporation of Brooklyn, NY. (PT Boats Inc.)*

Figure 3-23. *An inside look at the crews quarters looking forward. Here you get a good look at the crews lockers with air holes, and the roomy bunks. The door to the right led to the crews head. Notice the area has been painted with a bright white paint, that made it easier to see below decks. There are also over-head lights in this compartment for night use. (PT Boats Inc.)*

Figure 3-24. *A look inside the starboard side of the forward crews quarters looking aft. You can see the Mahogany drop down table for the crew meals. The water tight door leads into the boats Galley, with the crews bunk area being very comfortable with good mattresses. Flooring in this area was aluminum alloy, covered in an anti skid safety coating which resisted abrasion and wear. (PT Boats Inc.)*

Figure 3-25. *The day room cabin (looking aft) onboard the 103 class boats was considered at the time, to be that of a lounge area for the crew. It was pretty spacious and had six cabin windows, three to a side, for more air. It was located above the fuel tank compartment between watertight bulkheads, with a non watertight door to forward quarters, and sound-insulating door and hatch in after bulkhead to engine room. All of the furnishings above the day room floor were easily removable, so that flooring and tank side panels could be removed for inspection or service of the tanks. In the starboard forward corner, a hatch of at least 22 inches square was cut, with a hinge back top, to provide an exit from the day room, using a steel ladder with tubular rungs mounted on the house side. It had two couches that faced one another for comfort in mind. Potable water was carried in a single 200-gallon tank located in the bilge under the chart room. (PT Boats Inc.)*

Figure 3-26. *A look inside the boats engine room, showing one of three 4M-2500 Packard Marine Engines. This starboard side wing engine, was the battle station of the engineer on duty. A seat for the engineer is provided on the starboard engine from which he can start the engines, observe the instrument panel, control the spark and scoops, and shift clutches. To his right up high on the bulk head is the starting panel switches for starting the engines. The tractor levers in front of him are for shifting the engines into gear for forward, neutral and reverse. (PT Boats Inc.)*

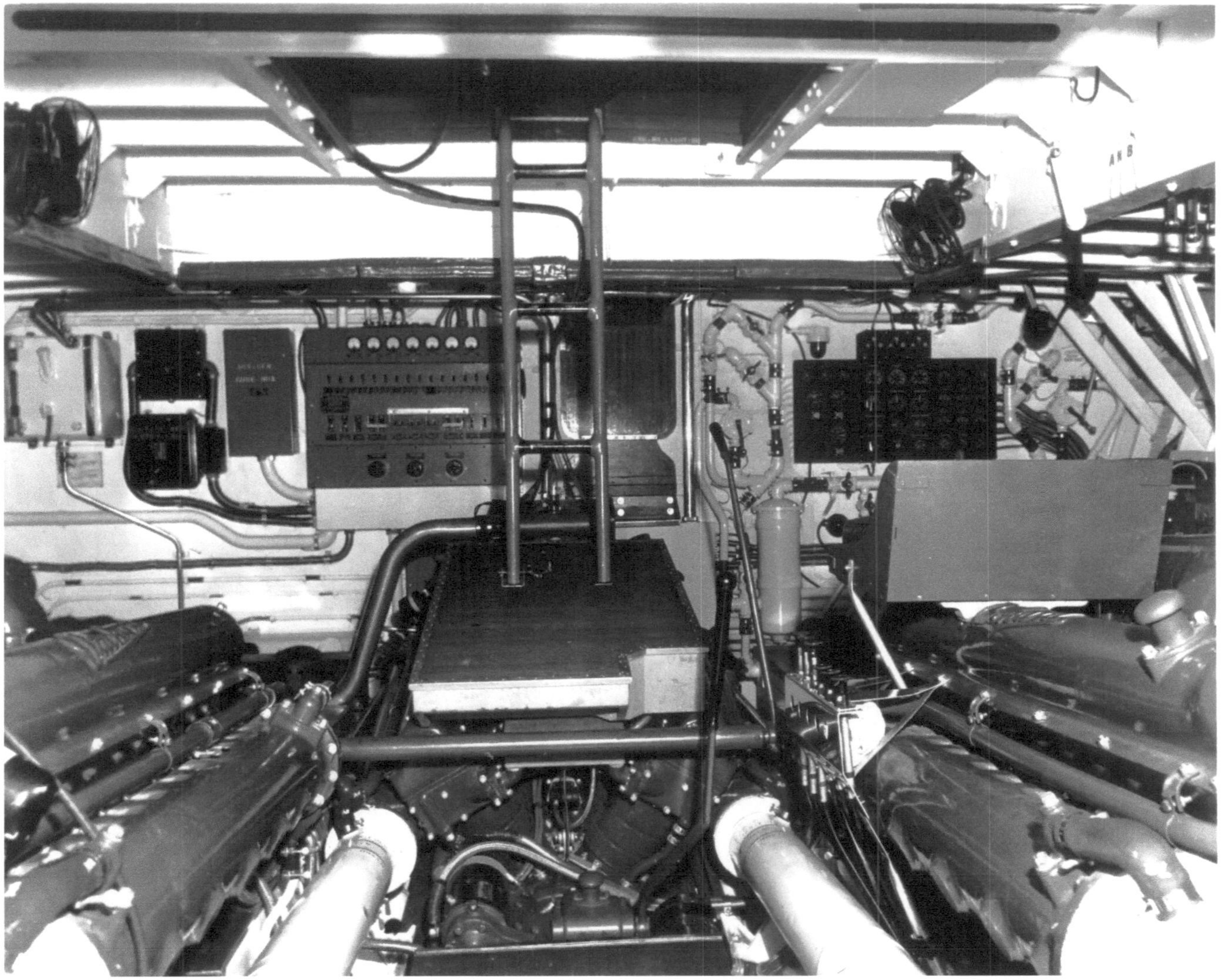

Figure 3-27. *A wonderful look into the boats engine room looking forward. We see the back of the engineers seat to starboard and to center, we see the ladder leading topside. The electrical system of the Packard Marine engine is similar to the electrical system used in automobiles as is used an electrical starter, generator, generator regulator, and storage battery. The panel to the left of the ladder shows the main electrical panel, which showed battery ammeters, engine room load switch, battery circuit breaker switches and other electrical instruments which the engineer maintained. To the right of the ladder is the main instrument panel for oil and water temperature gauges, engine annunciator's which relayed the captains signals for the engine shifting levers, RPM and Manifold pressure gauges. (PT Boats Inc.)*

Figure 3-28. *Starboard side look at the engine room looking aft. Just forward of the engineers seat is the boats auxiliary generator. This generator was a very vital piece of equipment, as it would run the boats cook top, refrigerator, coffee maker, and chart-house that needed this reliable source of current to run equipment. This gasoline powered generator would be used, when the engines were not running. Great care was needed to insure it would be running when needed. White pipes are exhaust which lead out to the boats transom. (PT Boats Inc.)*

Figure 3-29. *Engine room main battery panel. The engineer in the engine room would make sure that batteries were producing enough current for items such as the galley, port and starboard batteries, main engine batteries, aux generators, main engine aux generators and more. (PT Boats Inc.)*

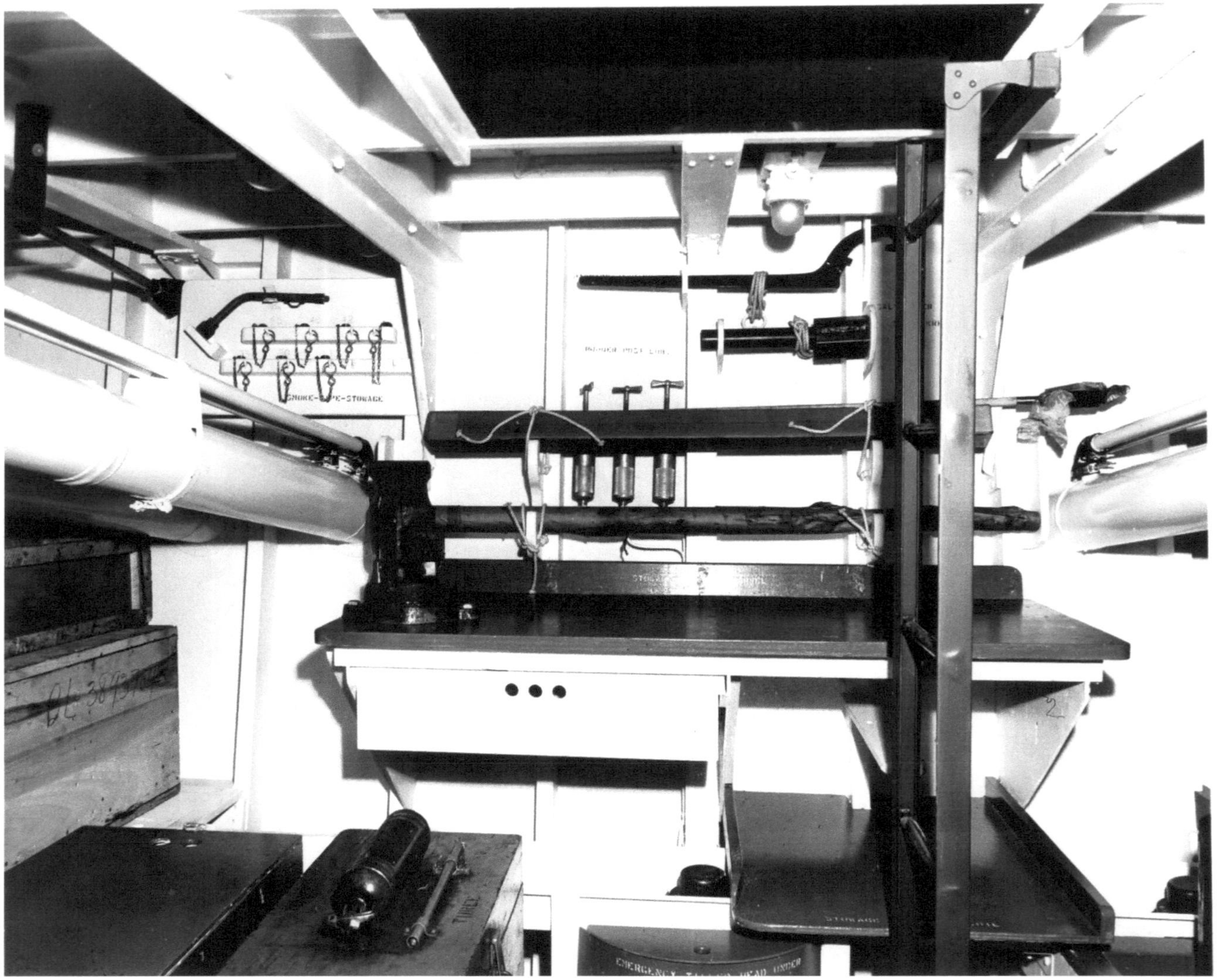

Figure 3-30. *This was the area known as the Lazarette. It was the last compartment at the stern of the boat, and was originally designed for a work space. It was accessible through a hinged deck hatch with metal ladder to floor, and through a watertight door in after bulkhead of engine room. In this space was the auxiliary steering apparatus, where if need be they could rig an emergency tiller. It would become a perfect spot for storing gear, such as oil and paint cans, lubrication gear, smoke generator parts, and even extra ammunition. Notice the small work bench area, exhaust pipes and ladder to topside. (PT Boats Inc.)*

Figure 3-31. *The M-2 Aircraft 50 caliber machine gun was used on PT boats. The M-2 had a shorter barrel and weighed less then the standard 50 caliber. It had a higher rate of fire as well, although a shorter range. The basic aircraft gun came without a trigger or the side bracket with the charging handle called the retracting slide group (RSG). PT boats used this basic gun along with a mount supplied trigger and usually a simple bolt handle for charging the gun instead of the RSG. Unlike the RSG, which operates against a short stud and remains stationary during firing, the bolt handle travels back and forth with the bolt. On the MK 17 mount, the upper and lower triggers operated both guns. The boats had two gun tubes carrying twin 50 caliber machine guns with, one port side mid ships attached to day room cabin, and one starboard side opposite the helm. This photo shows the guns that were on the port side of the boat. (Frank J. Andruss Sr.)*

Figure 3-32. *The standard weapon that was added to the 103 class boats was the 20mm Oerlikon Gun. The gun Swiss in its design, was manufactured originally by the Oerlikon Machine tool works. This gun was a close range, high angle, anti aircraft gun, which was singly mounted on a pedestal. The gun could be trained through 360 degrees. The gun could be fired automatically, as long as the trigger was pressed, and their was ammunition in the magazine. This weapon carried a cylindrical drum which held 60 rounds of tracer and high explosive rounds. The mount was a heavy Mark IV mount which was actually used on destroyers. Later the mount would be removed in favor of a light weight mount, which saved a good deal of extra weight. The gun was mounted on the stern of the 103 class boats coming out of the factory. (Frank J. Andruss Sr.)*

Figure 3-33. *The helm area of the 103 class boats. The three throttles to the left were controlled by the Captain, and when engaged in forward, neutral, or reverse, it registered with the engineer, in the engine room. In that way he knew in what position he would shift the engines. To the right of the throttle box is a long piece of metal, which was the speed control lever. The center or main control panel shows the three top rpm gauges, while the bottom three were the manifold pressure gauges. Around the main panel are plunger buttons, and switches that controlled engine room signal buzzer, stern lights, navigational horn, main control switch, aft signal, lock out signal switch, and more. The hinged box to the right of the control panel when opened contained the switches for launching the four torpedoes. Notice the metal wheel, which on the top has a suicide knob built in, making turning the boat much easier. (PT Boats Inc.)*

Figure 3-34. *Mr. and Mrs. John D. Bulkeley on the launching stage. Mr. George Hicks from the National Broadcast Company conducted the on air ceremonies, with music provided by the Richmond County American Legion Band of Staten Island. The Master of ceremonies was Henry R. Sutphen Executive Vice President of Elco, and the speakers were Govenor Charles A. Edison, Lt. John D. Bulkeley, Mayor James J. Donovan, and Rear Admiral Adolphus Andrews. Moving pictures were taken for the News reels. (Glenville Tremaine)*

Figure 3-35. *With much delight and excitement on Saturday May 16, 1942, Elco began their grand ceremony around 4:30pm, with the launching of PT-103. This would be the companies first eighty-foot Elco designed boat coming off the assembly line. With the boat hooked up to the dock side crane, it would be christened by Mrs. John D. Bulkeley. Here Lieutenant Bulkeley addresses the hundreds of workers and guests that have turned out for this very special day. He would tell the workers that he has seen first hand the production in action and tells them "Well Done". Bulkeley tells the workers that in a sense the placing of a torpedo into an enemy ship, is a final culmination of our combined efforts, the efforts of each of you workers. (Glenville Tremaine)*

Figure 3-36. *PT-103 has been hooked into the cradle. She is slowly lowered into the boat basin by the dockside crane. It will be the first time she will be floating on her newly constructed hull. (Irwin Chase)*

Figure 3-37. *With hundreds of workers and guests in attendance, PT-103 sits in the boat basin. On the ceremony stand with backs to the photo, is Executive Vice President of Elco Henry R. Sutphen, and to his left is the sponsor Mrs. John D. Bulkeley with bouquet of flowers. She is now floating on her hull for the first time, as the engine crew is beginning to start up her three engines. This would be a momentous occasion for Elco, as this new designed boat, would be just the start for more eighty-footers to come off the assembly line. (Irwin Chase.)*

Figure 3-38. *PT-103 at the dock, shortly after her launching. Notice on the port side the boats search light which was a Ray Line, exclusive design by Elco. They were powerful lights that could spot both hazards and enemy craft in the darkness. These class boats used the lights made by the Portable Light Company of New York. To the left is Lieut. Frank W. Rowe, Assistant Supervisor of Ship Building for Elco, Captain George A. Alexander, and Elco's Executive Vice President Henry R. Sutphen. (Irwin Chase)*

Figure 3-39. *Another look at the newly launched PT-103 at the main boat basin. The officer on the deck looking away from the camera is Lt. Cmdr. Frank W. Rowe, Elco's Assistant Supervisor of Shipbuilding, inside the helm with leather skull cap on is Managing Constructor of Elco, Irwin Chase. Others in the helm are engineers that will begin testing all components. Behind Chase is Robert Montgomery, who served with Squadron 5 commanding PT-105, Later he would become a famous actor, having the lead in the movie "They were Expendable" about the exploits of PT Boats in the Philippines with Motor Torpedo Boat Squadron 3, portraying Lt. John D. Bulkeley. (Irwin Chase)*

Figure 3-40. *A port-side side run showing PT-103 in all her glory. Her beautiful lines are unmistakable, and she looks majestic gliding across the water. It is easily mistaken when looking at her to think she was made for pleasure, when in reality she was built for war. At this point the boat is still in Elco hands, as the sign reads run number 29. The men on the boat are part of the Elco trial crew and will test all of the boats components, making sure the boat is 100 per cent before turning her over to the US Navy. The Navy however will run their own tests as well, before shipping the boat to her operating squadron. (Frank W. Rowe)*

Figure 3-41. *A wonderful hand built model of PT-103, constructed by Hull Department employee Donald Rosencrantz. Elco had him build this for the special launching ceremonies, and was on display inside the entrance to the main office. Rosencrantz would build several models for the company over the years, using a combination of wood and white metal parts. (PT Boats Inc.)*

Figure 3-42. *With the success of the eighty-foot design, Elco would continue to turn out these boats, one every 60 hours. After being towed out of the assembly building, they would make their way to the dockside crane, being gently lowered into the basin. To make room in the assembly building, the boats would be towed to the outdoor boat shed, where more assembly would take place. Here one of the boats is in the outdoor boat shed where it will undergo more assembly. Once again she will be towed, this time to building 16, the wet boat basin for many of the final assemblies. (PT Boats Inc.)*

Figure 3-43. *An outside look at building 16, the wet boat basin. Up to six boats could be docked here for many of the boats final assemblies. The doors could be opened in the summer months, and closed in the winter as this building was heated. The boats typically needed to "soak" or remain submerged in water for several days to allow the wood to fully absorb moisture and expand, creating a tight seal between the planks and preventing leaks. (Frank J. Andruss Sr.)*

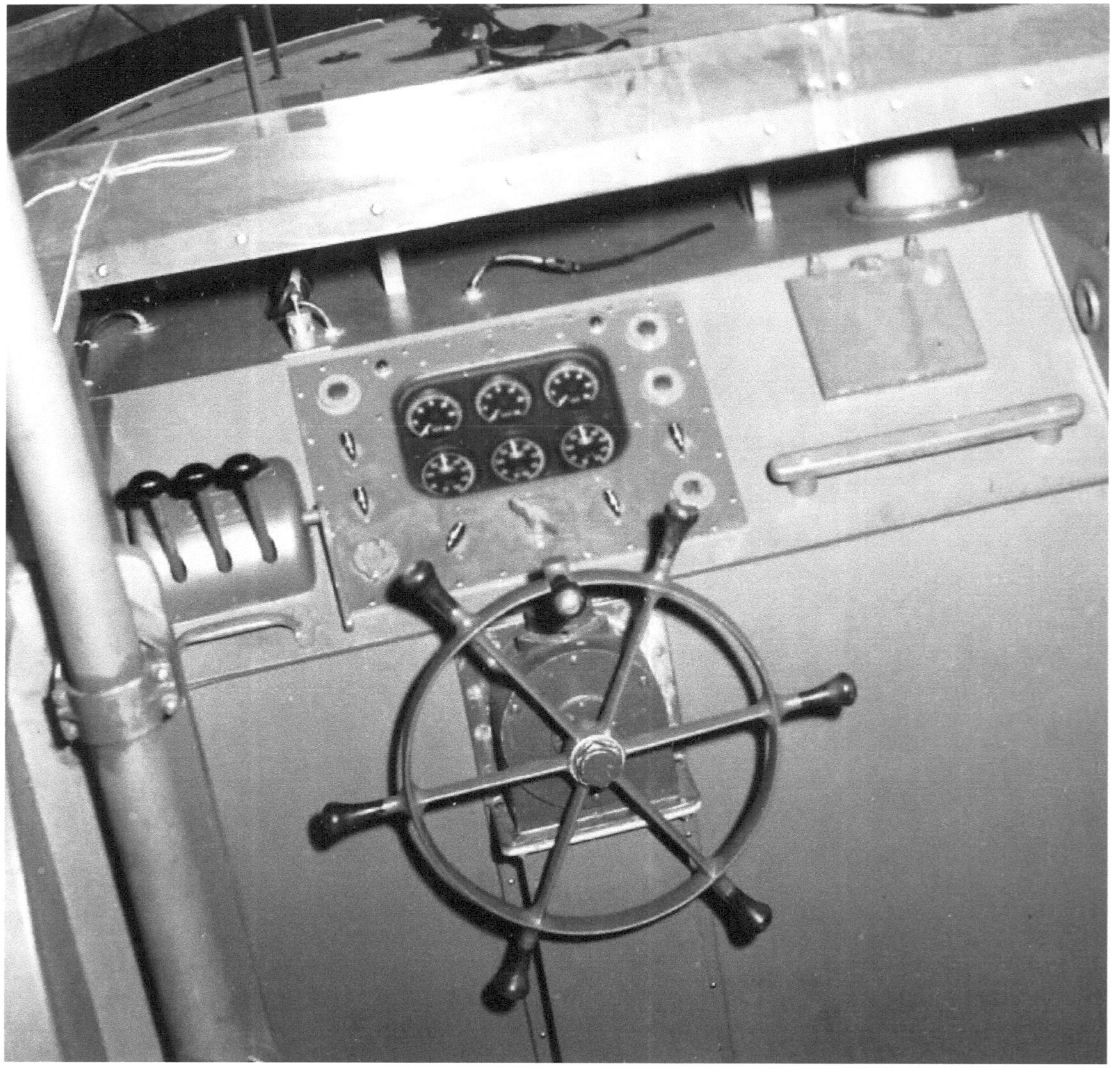

Figure 3-44. *An inside look at the wet boat basin showing one of the eighty-footers. You get a good look at her helm area, and how simplified her bridge controls were. To the right is the door leading to the boats chart-house and below deck crews quarters. The aluminum pole to the left is for the boats search light. (PT Boats Inc.)*

Figure 3-45. *This boat inside the wet boat basin will continue with her assembly of weapons systems, ammo boxes, torpedo racks, and other sub assembly work. We see a good look at the starboard side 50 caliber gun turret, as the guns will be checked here for clearances. The rails around the turret, are gun depressor rails which were installed, so that gunners could not shoot into their own boat. All systems will be checked here, and then re-checked. (PT Boats Inc.)*

Figure 3-46. *Now that the boats have left the wet boat basin for fitting out, they would be towed to the fuel barge. Elco did not have any underground gasoline tanks on her property. They had this fuel barge from Gulf Marine which was inside the boat basin, and fueled the boats as needed. Here PT-175 is backed up to the barge. We are seeing the tail end of the fuel barge with a floating dock tide behind her. It's not a big barge by today's standards, with this one being 30'x 90/120'. The house on the barge is for pumps and generators, which is used for self propulsion. (Frank J. Andruss Sr.)*

Figure 3-47. *The tractor at Elco was an Austin Western, 4-44 model that featured all wheel drive, and was an all wheel steer tractor. They were manufactured in Aurora, Illinois, and were used at Elco for removing completed boats from the main assembly building and moving them to other areas of assembly and to the dockside crane for launching. (Naval War College Archives)*

Figure 3-48. *A nice beautiful port-side look of the PT-117. This was another 103 class eighty-footer. Her lines are unmistakable and offer a look into a design that was second to none. Still in Elco hands she will like sisters, undergo much testing to make sure there are no errors, before being turned over to the Navy. (Irwin Chase)*

Figure 3-49. *As the war progressed reports from the forward areas were always sending back information to Elco. This was a very important aspect of development, as boat crews made suggestions as to armament, torpedoes, and other areas in which the boats could be improved. As Elco took these suggestions and studied their worth, changes in the newer boats were taking place. As the 103 class boats, numbered to PT-196 came to a close, sub classes of the boats were being being upgraded. Here is a photo of one of those upgrades as a 20mm Orelikon cannon has been added to the bow. She is still in Elco hands in this photo. (PT Boats Inc.)*

Figure 3-50. *Several of the Elco 80 foot boats would make their way to the Motor Torpedo Boat Squadrons Training Center, serving with Squadron 4. Here PT-141 shows her forward torpedoes in place, but her aft torpedoes have been substituted for Depth charges. She still retains her factory installed 20mm cannon on the stern, and her bow is still empty of later armament upgrades. (Charles B. Jones)*

Figure 3-51. *The dockside crane gets set to gently place another boat into the boat basin. This time it is PT-131. Miss Marion Sutphen and Lt. Cmdr John D. Bulkeley would christen the boat in a ceremony held at the plant. The boat would go on to serve with Squadron 7 in the New Guinea area of the Pacific. This squadron was awarded the Navy Unit Commendation for action in the New Guinea area from April 1, 1944, to February 1, 1945. (Irwin Chase)*

Figure 3-52. *A great look at PT boats coming and going from the Elco Yacht boat basin. The boat coming into the basin is PT-345. As you can see her mast is in the stowed position, her forward gun on the bow has not yet been assembled onto the mount, her machine gun turrets are covered, as well as the stern mounted 20mm cannon. Notice the traffic of PT boats that are operating in Newark Bay. (PT Boats Inc.)*

Figure 3-53. *Port side view of PT-377 as she leaves the boat basin at Elco. Notice she carries a forward 20mm cannon on her bow, although she still has the old style MK-18 torpedo tubes. Later these tubes would be replaced by the roll off racks and the newer MK-III torpedoes. Looking forward of the two aft ventilators, her engine room hatch has been completely redesigned to make access easier and allow for better ventilation below deck. (PT Boats Inc.)*

Figure 3-54. *Later war Elco showing them laying down the decking. A very good photo showing the workers at Elco with an outside section of Plywood Decking about to lay it down on the decking frames. This section runs just about the full length of the Boat (allowing for the center sections). This would replace the planked decking that could be found on the earlier boats. Elco needed to stop using mahogany (especially the preferred African) since they were always having problems obtaining it. Also, plywood decking was one of primary original designs on the experimental boats. Only 80" Elco's PT-103-196, 314-367, and 372-383 had planked decks, later boats had plywood decks. (Frank J. Andruss Sr.)*

Figure 3-55. *Eight PT boats that will eventually form Squadron 25 have assembled in the Yacht boat basin at Elco. Several of these boats have their forward MK-18 torpedo tubes and depth charges at the stern. All but two of the boats have their SO-Radar masts in the lowered position. Many of the boats will have radar installed which gave them a distinct advantage over the enemy. Early boats did not have radar installed. Some of the boats also have their torpedo tube covers on to protect the tube from the salt water spray, which others have canvas attached to these openings. The 20mm cannons on their bows have not yet been added to the mounts. (PT Boats Inc.)*

Figure 3-56. *PT-556 has been pulled into the outdoor boat shed. Looking as though she is almost completed, the boat is awaiting her MK-XIII torpedoes. Canvas covers are added to the boats two gun turrets. Note that the radar mast has been folded down. Masts on Elco boats folded back to the center line, at a 70 degree angle to clear the after fifty caliber turret. The boat would serve with Squadron 29 in the Mediterranean. (Naval War College Archives)*

Figure 3-57. *PT-731 has been attached to the twin boom crane and is swung out over the basin. The 731-760 boats were built expressly for the Soviets. Only the lead boat (PT-731) was actually built at Elco, the remainder being shipped as "kits" for the Soviets to assemble. (PT Boats Inc.)*

Figure 3-58. *PT-731 has now been placed into the boat basin. Here you can clearly see her MK-19 torpedo tubes. These tubes were fired using compressed air and were fitted with an internal heater to deal with the cold climate of the Soviet Union. From here she will be towed to the wet boat basin for the next phase of her assembly. (PT Boats Inc.)*

Figure 3-59. *Here we see the Elco work boat, nicknamed "Pee Tee" in the main boat basin. This boat would be used to tow the PT boats to areas of assembly and to the fuel barge. It would also be used to tow in barges and other supplies that were off-loaded by ship outside of the plant. It would continue to operate after the war until the plant closed in 1949. (PT Boats Inc.)*

Figure 3-60. *This photo was taken in the winter months, but weather did not stop the company from their continued efforts, in producing the PT boats. Underway trials continued in the winter, a cold but necessary step in getting the boats to their intended squadrons. To the man's left just below his hand, and (on the starboard bulkhead) of the day-room cabin, is what was known as a Lux Pull Box. It was used for the CO2 Fire Extinguishing system. The CO2 tanks were installed in the same location with manual releases on the port side aft of the bridge and outside of the engine room hatch. No automatic discharge for the fire extinguisher system was provided. Potable water was carried in a single 200-gallon tank located in the bilge under the chart house. Just underneath the cabin window was a window drain scoop. The raised apparatus on the deck was a mushroom vent, which was located port and starboard over the boats wing gasoline tanks, and two more located in the area over the rudder compartment. They were used to allow for ventilation, air circulation and air exchange. (PT Boats Inc.)*

Figure 3-61. *Beautiful port side look at PT-543. She is just outside the Elco break water and has been undergoing testing. She carries four MK-XIII torpedo tubes in roll off racks and a 20mm cannon on her bow. In this photo she carries a regular mast, but that will be updated to carry the SO Radar mast. Not yet added is the boats 37mm auto cannon, and 40mm Bofor's gun. After being turned over to the US Navy, she would serve with Squadron 37 in the Pacific. (Glenville Tremaine)*

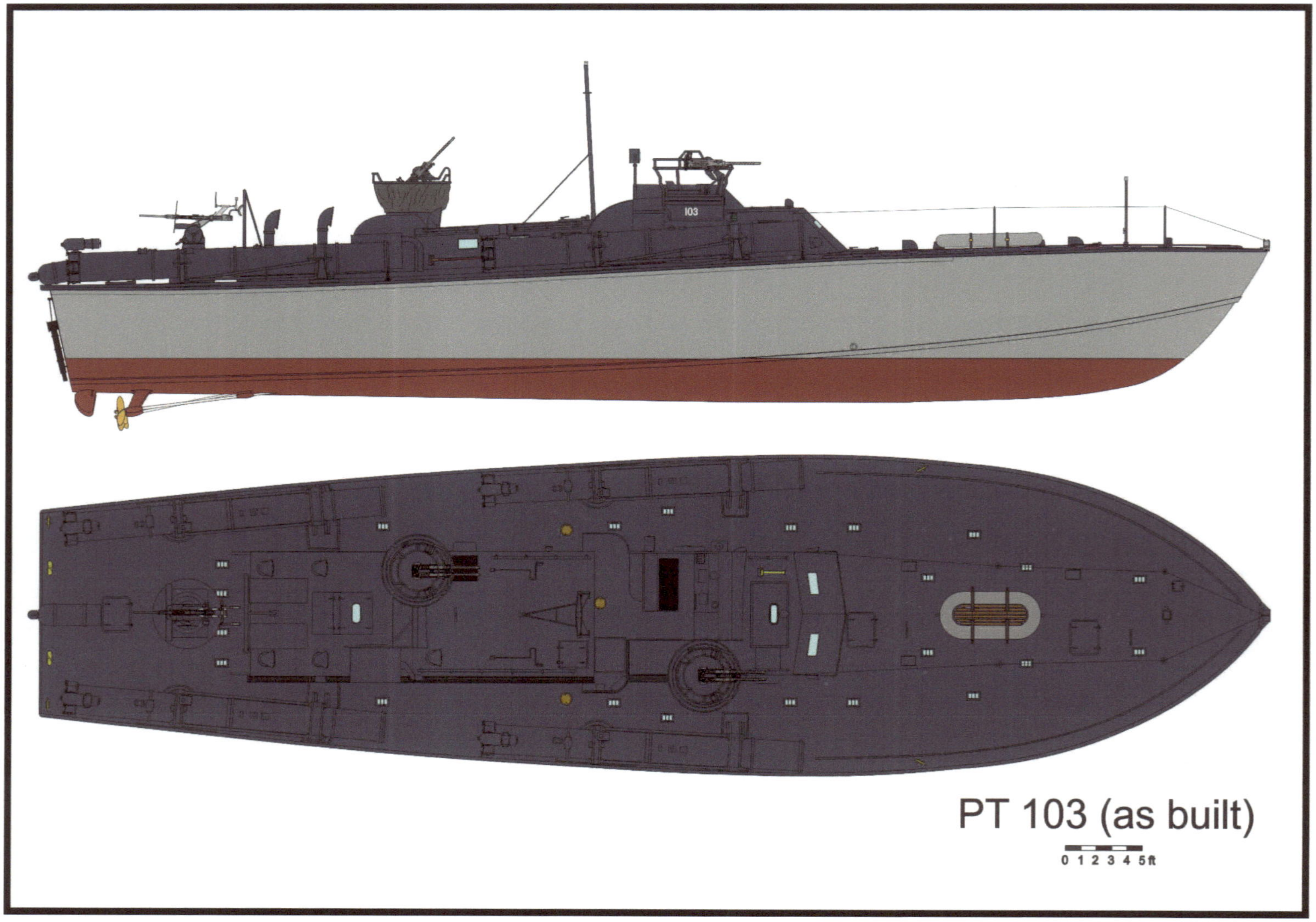

Figure 3-62. *Wonderful artist rendition of Elco's first assembled 80 foot boat, PT-103. This is how the boat looked and was painted coming off the assembly line. This class ran to PT-196. (Al Ross Drawing)*

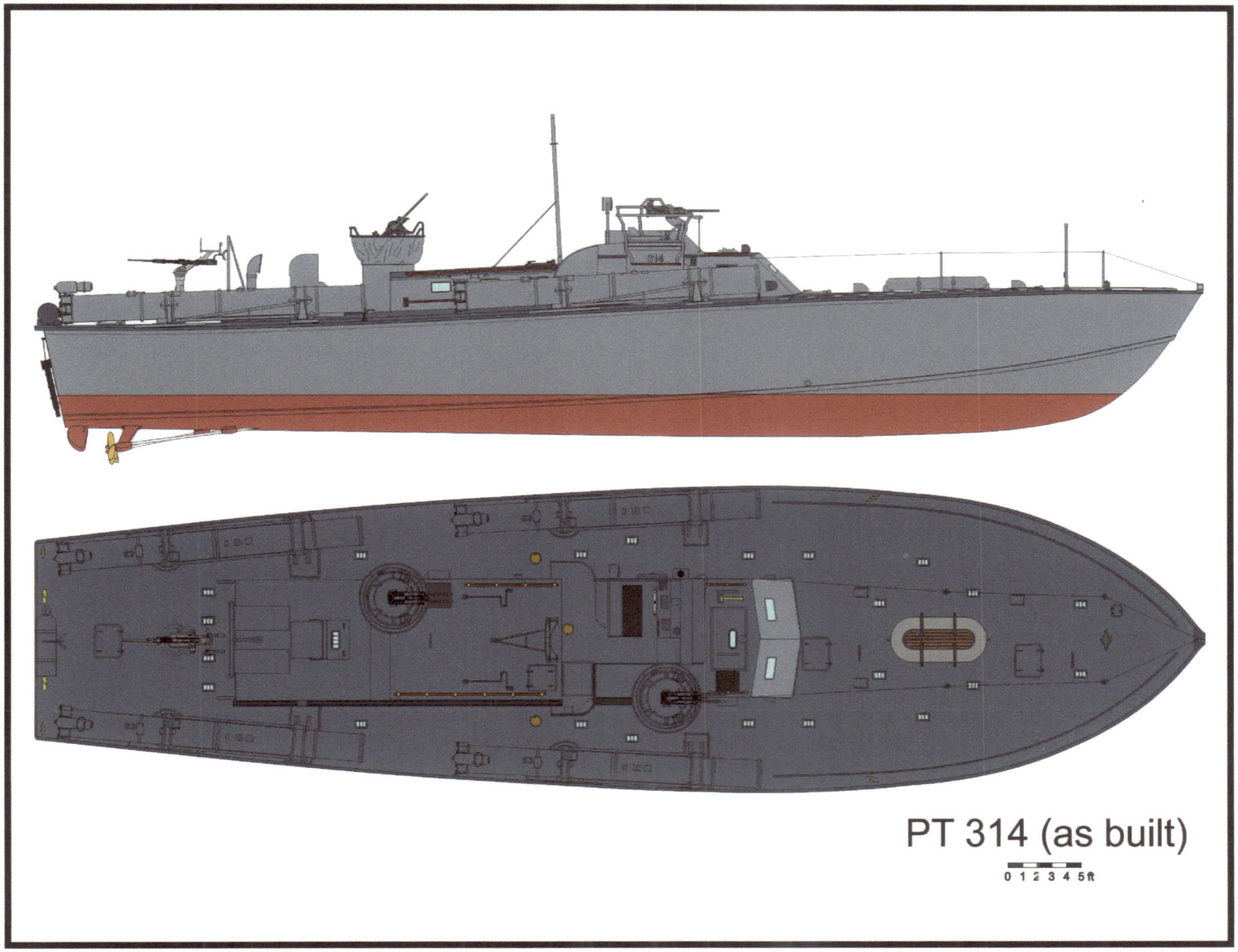

Figure 3-63. *The next class of 80 foot boat began with the number PT-314-367 as this artist rendition depicts these next class boats in the colors coming out of the assembly building. One of the changes to the boats was the redesigned engine room hatch, which had easier access and better air flow. Vents around the hatch were also re-positioned, and the heavy MK-4 pedestal mount for the stern 20mm cannon has been updated to the lightweight MK-14 mount. (Al Ross Drawing)*

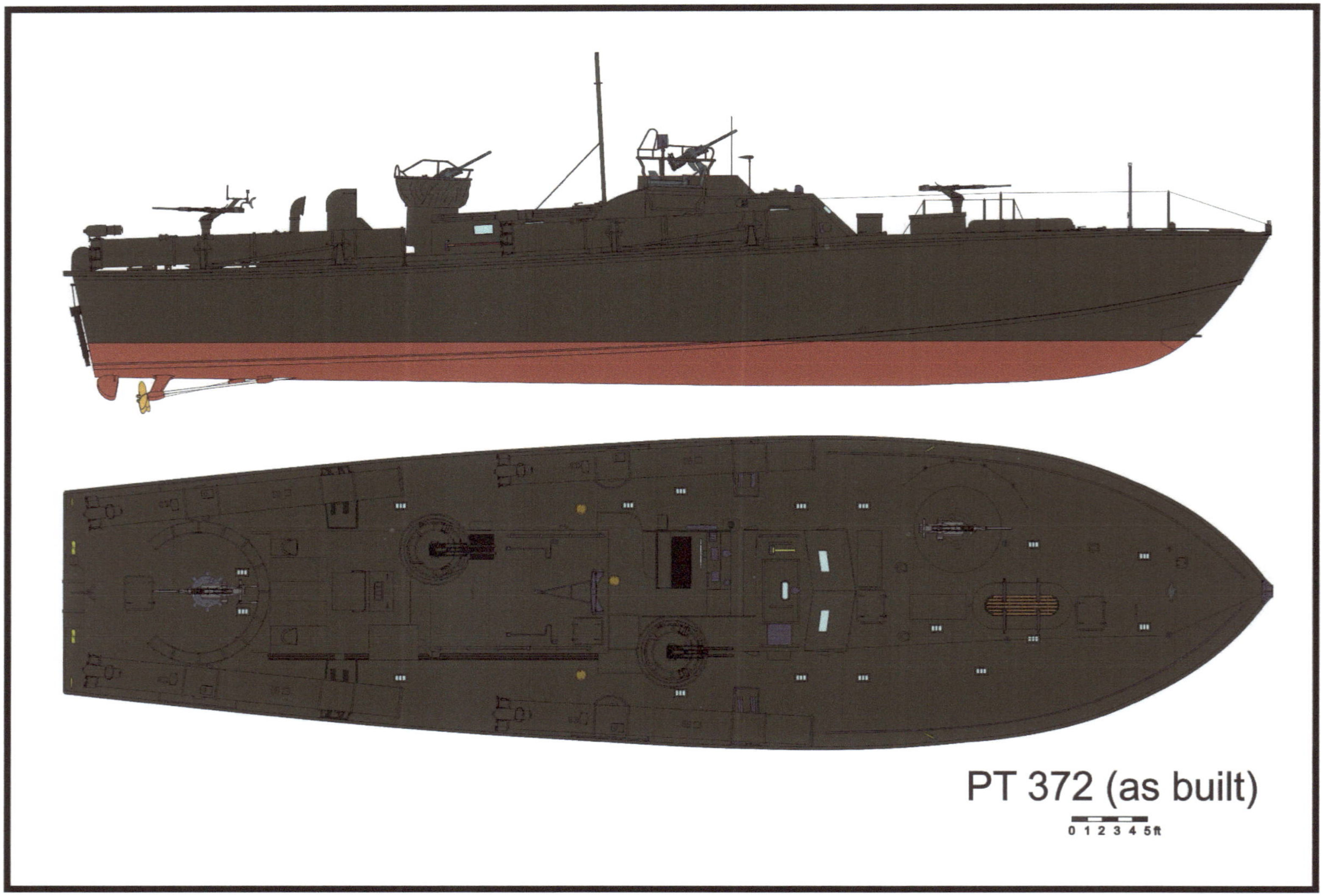

Figure 3-64. *The next 80 foot class of boat would be PT-373-383. This again depicts the boat as built at Elco coming off the assembly line. Notice at this point the two toned gray colors have now been replaced with the preferred green tones, and a 20mm cannon, with MK-14 mount has been added to the port side bow. (Al Ross Drawing)*

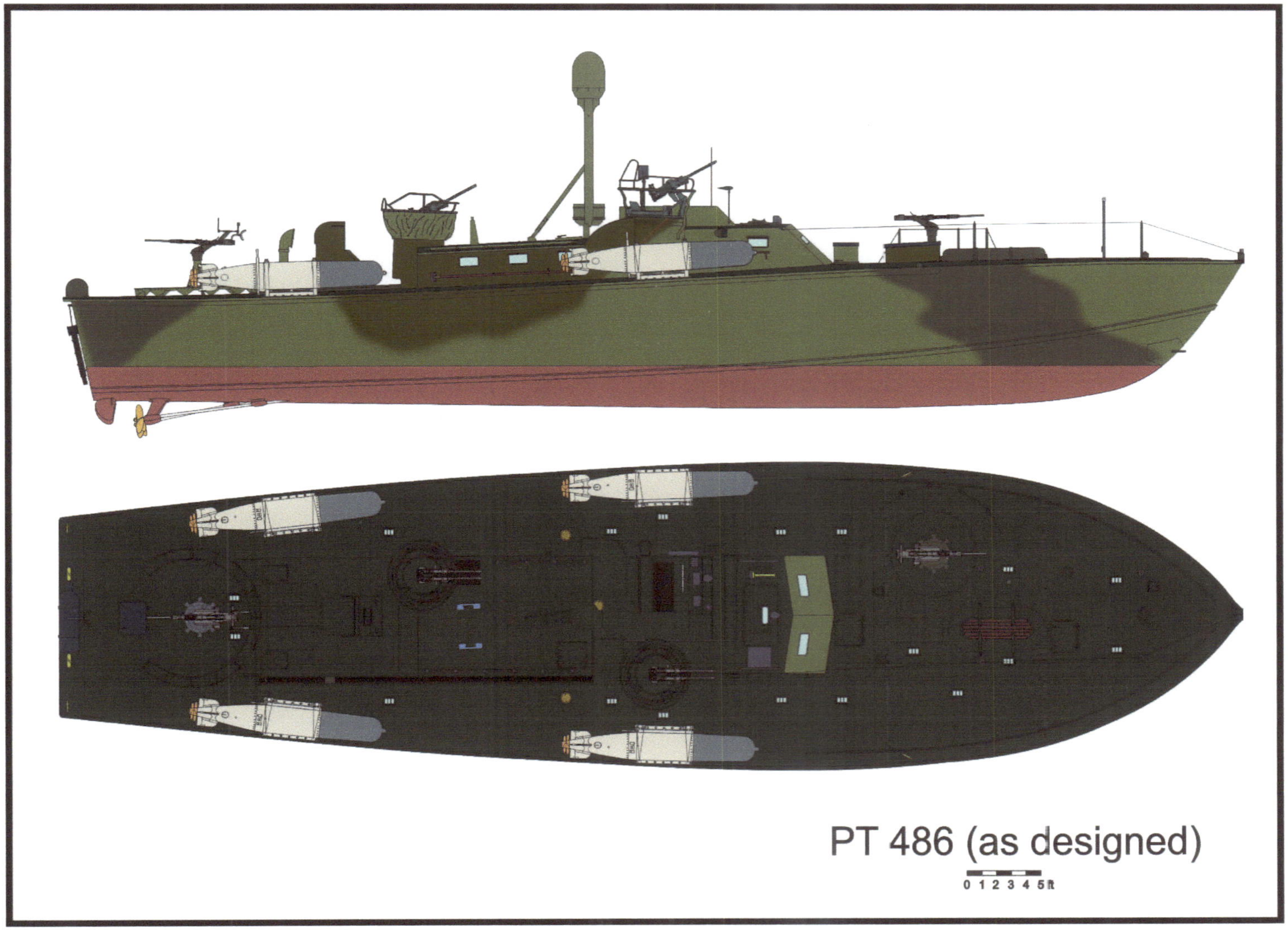

Figure 3-65. *The next class 80 foot boats would be PT-486-563. This drawing once again depicts the boats as they looked coming off the assembly line. Notice these were the first boats being painted using a camouflaged color, and they have the upgraded MK-XIII torpedoes and roll off racks, as well as the factory added SO-radar. (Al Ross Drawing)*

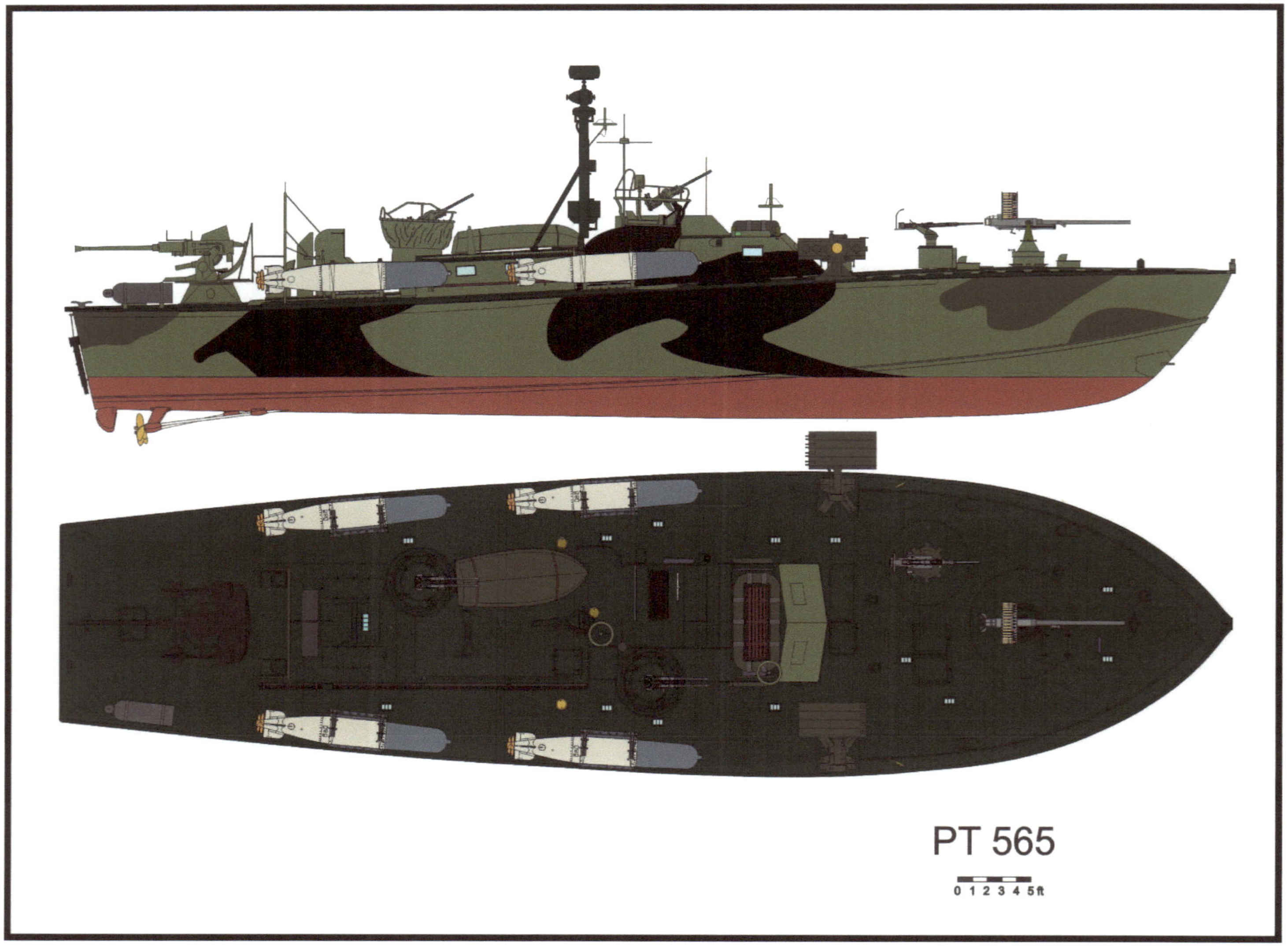

Figure 3-66. *The next class of 80 foot boats would be PT-565-624. One can clearly see the changes topside with these boats coming out of assembly. They carry a 37mm auto cannon on the bow, as well as a 20mm cannon. On the stern is the factory installed 40mm single Bofor's gun. She also carries the type 50 Rocket Launcher, that were used on the PT BOATS for added fire-power. Installed is the most up to date radar system to be used the SO-3. It should be noted that PT-623 and PT-624 construction was canceled in September of 1945. (Al Ross Drawing)*

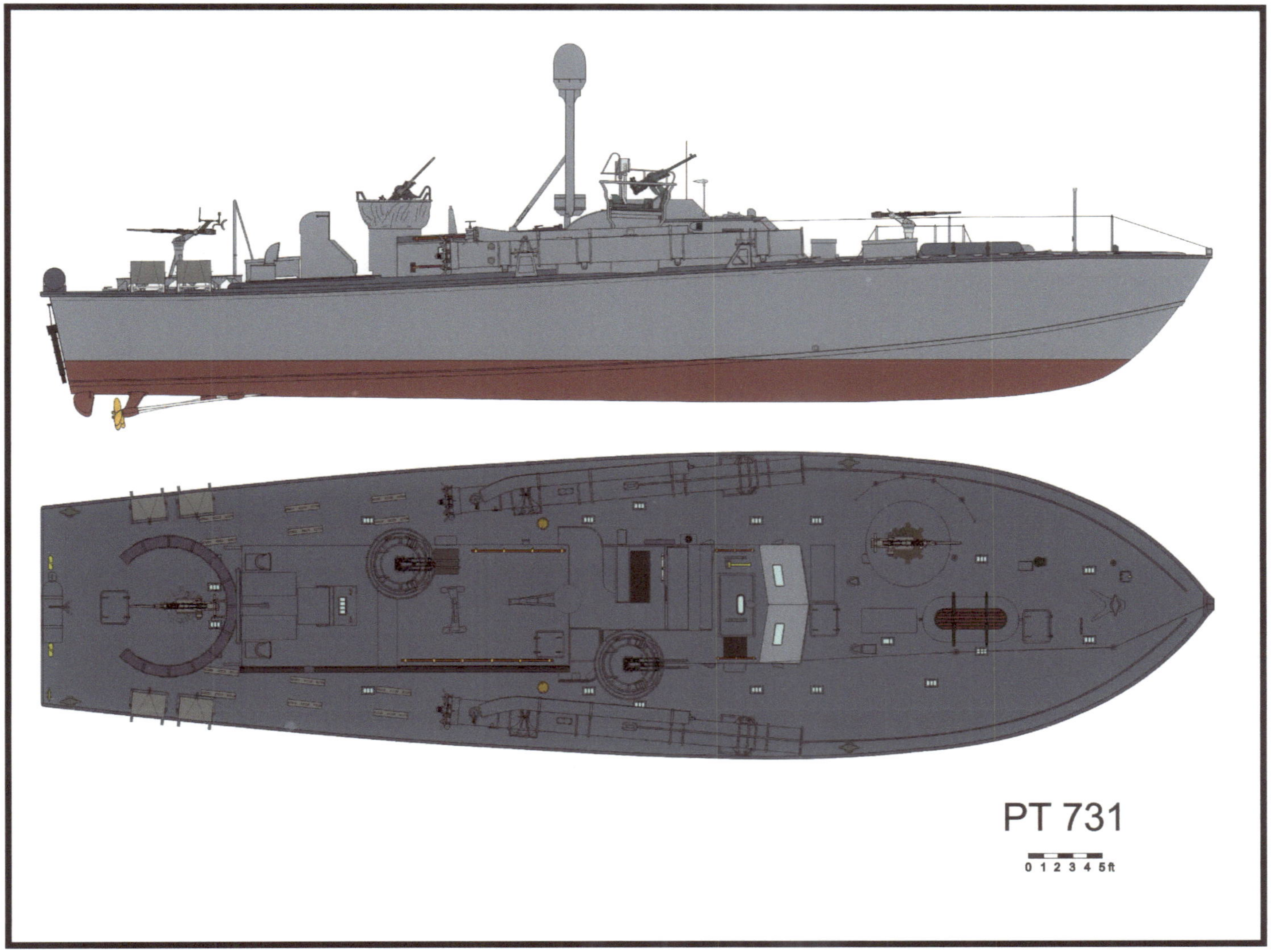

Figure 3-67. *This last class of 80 footers were built as knock down kits for shipment and assembly in the Soviet Union, and numbered PT-731-790. It should be noted that these boats were built to Soviet specs and included the MK-19 torpedo tubes, which were heated and needed in the cold climate in Russia. They also carry the earlier So-Radar. PT-761-790 had their construction canceled in August 1945. (Al Ross Drawing)*

Chapter 4

The Workers That Made It Happen

It would have been impossible to build Uncle Sam's PT boat Navy, without the dedicated and skilled workers at Elco. This American boat building and electric motor company operated from 1893 and would close for good in 1949. Elco was the front runner in building pleasure boats before the war, so their experience in wooden boat building was paramount.

Elco signed a contract to build fifty motor launches on April 9, 1915, and by the first of May, they had the first boat's frames erected. This boat would be used to create standardized patterns for the remainder of the order. The speed with which Elco prepared the drawings and patterns strongly would suggest they were working on this design well in advance of the signings.

Elco received a telegram ordering five hundred additional ***Sutphens*** (their code name for the motor launches) and the supplemental contract was signed on July 9, 1915. Terms called for completion of 550 boats by November 15, 1916; excluding Sundays, Elco was being asked to build 550 eighty-foot boats in 501 working days, an unprecedented assignment. Elco fulfilled this twenty-two-million-dollar contract in 488 days, two weeks ahead of schedule. It was due to the hard work and dedication of the workers. They had incredible woodworking and boat building skills, that would serve the company well during the building of the PT boats. The period between the two World Wars was Elco's ***golden age***, for during these two decades the company built the cruising powerboats for which it is best remembered today. The world's most popular cruising powerboat was the Elco Cruisette, first drawn up by Irwin Chase in 1915.

The Depression hurt Elco, as it hurt most boat-builders, but the effects of the Depression didn't make themselves felt right away. Elco continued to build boats and pay for eight-page advertising spreads in boating magazines well after 1929. The low point came in 1933, when Elco was reduced to only three full-time employees: Irwin Chase, Glenville Tremaine, and general foreman Charles Lamont. To keep busy, they personally fulfilled orders for spare parts, answered correspondence, and ran exhaustive dynamometer tests on new engines, just for something to tinker with. Gradually, over several years, the old employees were rehired and boat-building resumed.

It should be noted that many of the experienced boat builders that were building pleasure craft before the war, stayed on with the company to build the PT boats. Workers looking for jobs came from all over New Jersey, New York and Long Island. They came by the Staten Island ferry, then caught a bus close to the plant, or walked if they lived close by. Many took their lunches to work with them, or if they were lucky, they could get one of the local boys to run to the local delicatessen for some good sandwiches. Elco had a small canteen on the grounds, should they wish to order something quick. The company provided the workers with company washed uniforms and aprons.

They were provided with all the necessary tools for boat building, although many of the carpenters that were there for a long time prior to the construction of PT boats brought in their own tool boxes and tools. With the rationing of gasoline, it was tough for those with cars to get back and forth to work. Elco with it's OPA award set it up for those workers with cars, where they could go to the gasoline station assigned exclusively for their workers. Car pooling was also quite popular during war time.

Elco also had what was known as an apprenticeship programs. This system was put into place on May 1, 1941, its purpose was to help in plant production and was a two year schedule of training, which was completed at the Vocational High School in Bayonne, with the actual mechanics work put into practice at the plant. Many who took these programs became full fledged mechanics inside the Elco shops.

At the height of its PT-boat production, Elco employed more than three thousand men and women working three shifts a day for six days a week. They produced, on average, one PT boat every sixty hours. This brought Elco six consecutive Navy "E" Awards for defense contracting excellence, and earned the company ten million dollars in profits on a cost-plus basis. Laying off some two thousand of these employees when the war work ended was the worst job Elco ever asked Glen Tremaine to do. In 1939 he had been promoted to works manager and had day-to-day charge of Elco plant operation, including PT-boat design, throughout the war.

Sutphen, Chase, and Tremaine scrambled to find virtually any kind of work to keep the remaining one thousand workmen gainfully employed. In addition to new boat designs, Elco used extensive woodworking equipment to produce the hardwood trim for special-edition Plymouth station wagons; they built bowling lanes and bowling pins; they even produced baseball bats field-tested by the Chicago Cubs during one spring training. It didn't matter; the war had killed Elco. In 1939 Sutphen had doubled the size of Elco's plant and tripled its capacity in order to build the PT boats, and now he had to carry that useless overhead. The market for pleasure yachts wasn't promising, since GIs returning from the war were more interested in raising families and resuming or starting careers. There wasn't much discretionary income for buying top-of-the-line cruisers.

Workers with seniority were kept on during the time after the war. Getting back into the pleasure boat building business was not easy with the huge over-head that Elco had. Elco's parent company management was no help. In 1947, John Jay Hopkins became president of Electric Boat, and he envisioned a defense contracting empire that would have made even Isaac Rice and Sir Trevor Dawson envious. Electric Boat had built most of the submarines bought by the Navy during World War II; its corporate branches included major shipbuilding as well as submarine construction, aircraft manufacturing, electronics and much more. Elco, still building wooden pleasure boats, was a low-tech dead-end as far as Hopkins was concerned.

He told Tremaine that the ten-million-dollar profit on PT boats was "water over the dam." The new motto of Electric Boat Company was **grow or die**, he said. In 1948 Hopkins gave Elco a year to grow; in April, 1949, the unhealthy limb was pruned. Tremaine was given the task of laying off the remaining employees—an even more sorrowful task than previously—and of organizing auctions to sell the plant, equipment, and inventory. On the last day of 1949, Elco's Works Manager, Glenville Sinclair Tremaine padlocked the gate of the Electric Launch Company, handed the keys to an Electric Boat Company security guard, and drove home.

Figure 4-1. *These four women workers are on their way to begin a shift. As more men were leaving Elco for the service, it was these young women that would pick up the slack in all of the departments. They are dressed for the times with saddle shoes and company overhauls. Notice behind them is the enclosed walkway which lead from Building 20 to the company vault. (PT Boats Inc.)*

Figure 4-2. *Office workers standing out front of Building 20. This was the main office area of the plant and included the offices of all management. Many of the women in this photo were personal secretaries, working for upper management teams. (PT Boats Inc.)*

Figure 4-3. *These women workers take time for a quick photo before their shift. They are posing in front of a 1941 Chrysler Crown Imperial, one of the several cars that Elco purchased for war time use. As with all United States built automobiles, production was suspended February 1942 due to World War II production demands, and did not resume until the 1946 model year. Executive Vice President Henry R. Sutphen used this car daily. (PT Boats Inc.)*

Figure 4-4. *A great photo showing the many women that made up the Electrical Department at Elco. Prior to World War II, factories in the United States were turning out automobiles, large and small appliances, and children's' toys. A mere month after the attack on Pearl Harbor, Hawaii, President Franklin D. Roosevelt ordered the establishment of the War Production Board. Its purpose was to convert the factories of peacetime industries into manufacturing plants for weapons and military equipment for the fight. These ladies were more then capable of working in this department which gave life to the little PT boats. (PT Boats Inc.)*

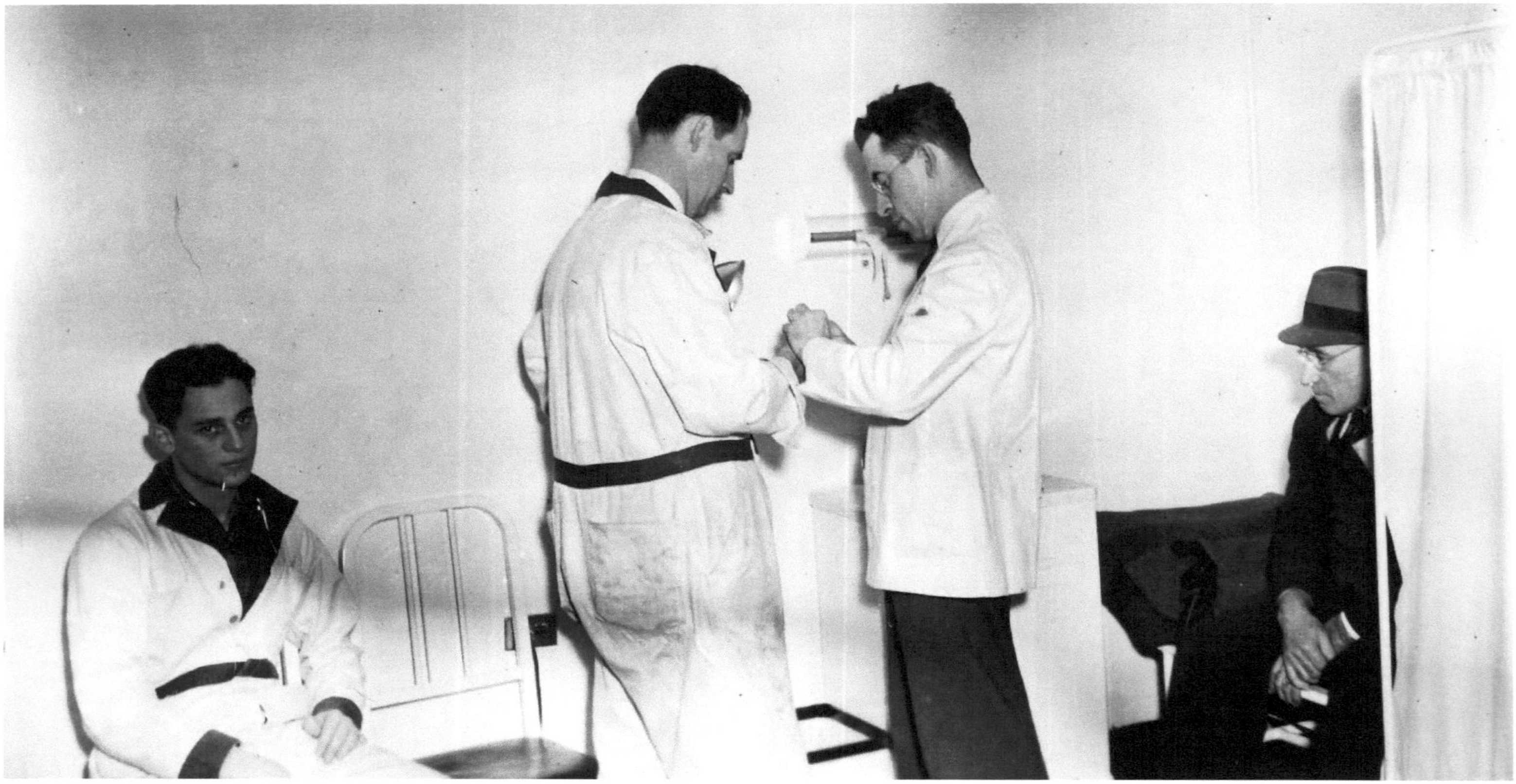

Figure 4-5. *The first aid office, located inside Building 20. Many injuries that occurred in a wooden boat building plant consisted of sprains, abrasions, wood slivers, and eye injuries. The head of this department was Mr. Alexander, seen second from the left. Most medical problems could be handled here, but for more serious injuries, Elco workers would be sent to Dr. William Wiren, who had an office at 8th Street and Blvd in Bayonne. (Irwin Chase)*

Figure 4-6. *An often forgot about area at Elco was the Photo Lab. Here they would process aerial photographs of the boats, filmed testing operations, documented accidents, photographed the assembly personnel at work, photographed complete boat series, and any other Elco photographs that were needed. Several photographers were on staff, including the famous Marine photographer Morris Rosenfeld, who took many photo's for the company. (Charles B. Jones.)*

Figure 4-7. *A very rare look at the photographers that made up the photo Lab at Elco. They would take the vast majority of factory photo's, and video that included assembly of the different boat classes in all of the plants departments. They also took the testing and underway photos of the boats. Here they are onboard PT-517 that is in the boat basin. (Irwin Chase)*

Figure 4-8. *Elco had their own Police Department which insured a high level of security in and around the Elco plant. They protected the facilities where boat materials were being produced. They patrolled other buildings, guarded gates, and enforced rules to ensure that the facilities were secure. Guards at Elco often received new forms of instruction in security, and many received new authority or were sworn in as special police officers. The man 3rd from left (white hat) is Glenville Tremaine, who oversaw this department as part of his regular duties. (PT Boats Inc.)*

Figure 4-9. *A look at the Sheet Metal Department workers. One might think with wooden boats, a sheet metal shop would not have a huge impact. By the 1940s, stainless steel and aluminum became more popular, cost-effective options. WWII helped take this trend to another level. The sheet metal shop was integral to producing prefabricated parts that were used in the construction of the boats. These parts were assembled with precision to meet the high standards required for wartime service. (PT Boats Inc.)*

Figure 4-10. *During World War II, hygiene in war plants was heavily emphasized through public health campaigns, mandatory hand washing, access to sanitation facilities like showers and toilets, and the use of disinfectants. Elco used posters and lectures to educate workers on proper hygiene practices like frequent hand washing. Regular medical checkups were often conducted to identify and address potential health issues. These lovely women workers are washing up in several hand washing stations at the plant. (Irwin Chase)*

Figure 4-11. *These officers have made the trip to Elco from the Motor Torpedo Boat Squadrons Training Center. A steady flow of officers and sailors alike visited daily to see first hand the operations of building the very boats they would take to combat. (PT Boats Inc.)*

Figure 4-12. *Navy Officers visiting the plant and learning first hand how metal products were being used on the boats. Elco felt it was very important that these officers learned how the boats were assembled, and could take the knowledge learned in an operating war time zone. (PT Boats Inc.)*

Figure 4-13. *A rare look into the Elco wet boat basin. Boats towed in here were in various stages of assembly. Three shifts of workers took care of the boats here for assembly of some weapons systems, ammo boxes, torpedo racks and other sub assemblies. Being brought here freed up space in the main assembly building, and allowed the wooden hulls to soak. Several boats you see here would later be placed with Squadron 4, based at Melville in Rhode Island, to be used as training boats for students. This area did have heat in the winter months. (PT Boats Inc.)*

Figure 4-14. *Elco employees formed their own in house Fire Department. They were all highly trained by the Bayonne Fire Department in the use of fire hoses, fire extinguishers, and other fire apparatus in case of a fire. As Elco was a wooden boat building plant, they took great precautions in fire safety, doing weekly training sessions with the Fire Department. (PT Boats Inc.)*

Figure 4-15. *Fire stations such as this one were strategically placed throughout the plant. They carried items such as fire extinguishers, axes, fire hoses, sand and more. Stations such as this one were instrumental in the case of fire, and could make the difference by having these emergency items on hand quickly.* (PT Boats Inc.)

Figure 4-16. *Here we see the welding department at Elco. Women workers are being trained on the proper techniques in soldering. The man in the white shirt is the welding foremen of the department. They are being shown how to heat with one hand and feeding filler in the other. Soldering is a process of joining two metal surfaces together using a filler called solder. The soldering process involves heating the surfaces to be joined and melting the solder, which is then allowed to cool and solidify, creating a strong and durable joint. (Charles B. Jones)*

Figure 4-17. *Workers in the main assembly area are working on the boats day room cabin, her forward chart house, and starboard side 50 caliber gun turret. The helm area has also been assembled here. The day room provided a space for crew to relax, but later would be used for berthing. The roof, outside panels and sides were made from mahogany plywood. The forward bulkhead panels, inside and out were made using birch plywood, with spruce frames with curved mahogany corners and covered on both sides with light plywood. The space between the frames in the superstructure is insulated with cork and the whole house is covered with airplane fabric, doped and painted. This will become PT-108 once her construction is completed. (Frank J. Andruss Sr.)*

Figure 4-18. *During the war years many slogans were made into posters and were seen just about everywhere. Here at the Packard Motor Car Company in Detroit, this Packard Marine Division employee holds up one of those signs. In back of him is one of the 4M-2500 Packard Marine engines used in the PT boats. (PT Boats Inc.)*

Figure 4-19. *Workers in a section of the joiner shop are busy measuring and fitting day room cabin and forward helm section of the boat. Employees that worked at Elco, during peacetime were woodworker's who made doors, windows, and decorative finish work, such as cornices and mantels, balustrades, and railings. Their skills were unmatched in the work they performed at Elco. (PT Boats Inc.)*

Figure 4-20. *Here in the main assembly section, these workers have one of the boats center self sealing gasoline tanks set up in the over-head lift. They will be placing it under the boats day-room cabin in the bilges. The Elco boats had three tanks, a port wing tank a center tank and a starboard wing tank. (Frank J. Andruss Sr.)*

Figure 4-21. *Elco did not have any underground gasoline tanks on her property, large enough to fill up the PT boats. They had this fuel barge from Gulf Marine which was inside the boat basin. We are seeing the tail end of the fuel barge with a floating dock tide behind her. It's not a big barge by today's standards, with this one being 30'x 90/120'. The house on the barge is for pumps and generators, which is used for self propulsion. To the right of the barge on the dock is a security officer and one of the security booth's over looking the water. (Irwin Chase.)*

Figure 4-22. *Elco Office employee Charles Waller (far right) was an executive assistant to the Works Manager, who at that time was Glenville Tremaine. He administered the issuance of initial authorization for production departments to proceed with all PT Boat construction, and conferred with reps from the Navy Department and Management. Charles had to study all contract specs and obtained data from the design agent. They are standing in front of PT-335. (Charles Waller)*

Figure 4-23. *Women workers on the hull are securing the planking. The fastenings to the main frames are Monel screws below the waterline and brass screws above and copper rivets or clinched nails were used for secondary fastenings. Due to the war requirements of copper, brass, and Monel, production completed after the summer of 1942, all fastenings were galvanized steel. (Irwin Chase)*

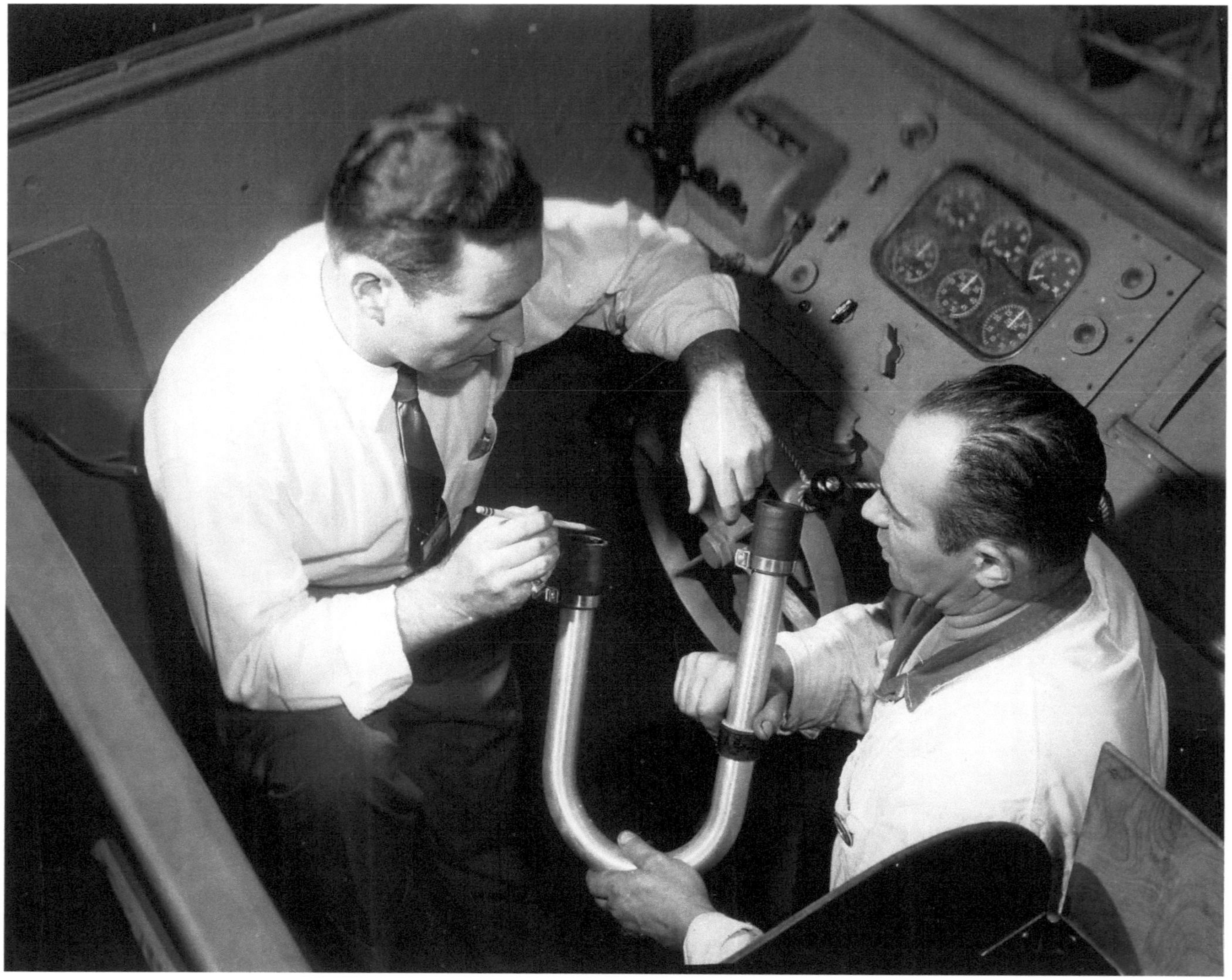

Figure 4-24. *Mechanical Engineer (left) and assembly worker are inside the helm of this 103 class Elco PT boat. They are looking over a piece of piping that will be connected inside the boats engine room. Fittings must be perfect so that no leaks develop while under the strain of combat. (PT Boats Inc.)*

Figure 4-25. *One of two Elco outdoor sheds which held a multitude of wood products used daily. The sheds were massive and also contained many other parts that were crucial to the assembly of the boats. These areas were high security as Elco police were stationed here at all times. Workers here had the responsibility to stock and maintain all boat parts and wood products which usually arrived by railway. (PT Boats Inc.)*

Figure 4-26. *Workers in the joiner shop are working on the inside of the boats day room cabin. Three workers are standing inside what will become the boats port side gun turret, which was attached to this cabin. This cabin was built of spruce frames with curved mahogany corners and covered on both sides with light plywood. The space between the frames in the superstructure was insulated with cork and the whole cabin covered with airplane fabric, doped and painted. (Irwin Chase)*

Figure 4-27. *Mahogany wood that has been laminated with a resin glue process is being locked together in a mold. From here it will go to the baking oven and heated to 140 degrees for two hours. New gluing techniques back then were able to withstand the salt air and water. They represented a great advancement in the building of wooden boats. (Naval War College Archives)*

Figure 4-28. *After being removed from the oven, the frame is removed. Here one of the workers is sawing the wood into two identical pieces. In this photo you can see the many lamination's, with the wood and glue bound together like steel. Planning rollers will bring the width down to precise specifications. (Naval War College Archives)*

Figure 4-29. *This worker is operating what was known as a pin router. This was a woodworking machine where the cutting bit is guided by a small, vertically adjustable "pin" that follows the contours of a template, allowing for precise copying of shapes onto a work piece, often used for creating decorative reliefs, custom-fitted recesses, and signs; essentially, it's a specialized router that uses a template with a guiding pin to create exact replicas of shapes. (Irwin Chase)*

Figure 4-30. *This worker is operating a 25 ton punch press. The punch comes down through the metal and meets with a matching die. Those larger flanged holes were done in the same manner. Those dimpled holes reduce weight and increase rigidity. The operator is using a foot peddle and engages a clutch on that flywheel to punch a quick clean hole. (PT Boats Inc.)*

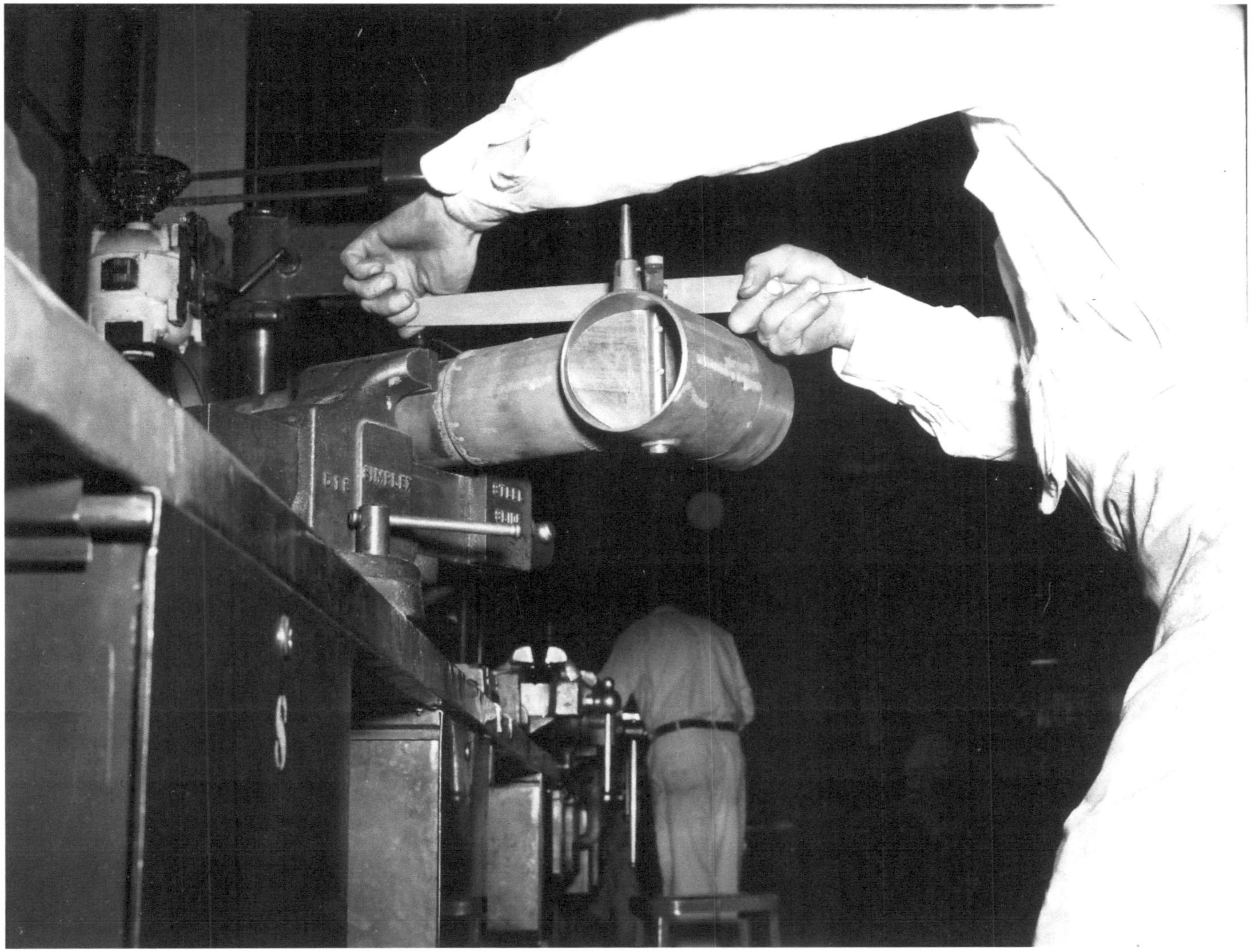

Figure 4-31. *One of the boats salt water outlet pipes can be seen here. These would be attached to the exhaust pipe outlets above the mufflers at the stern. They dumped into the inboard side of the exhaust outlet, around the outlet jacket, then into water jackets on the outboard side of the mufflers then into the mufflers. Mufflers were in the open position when the engines were started, and were open all the time above 1200 rmp. These outlet pipes were closed when the boat did not want to be detected and the exhaust was driven below the surface of the water. (Irwin Chase)*

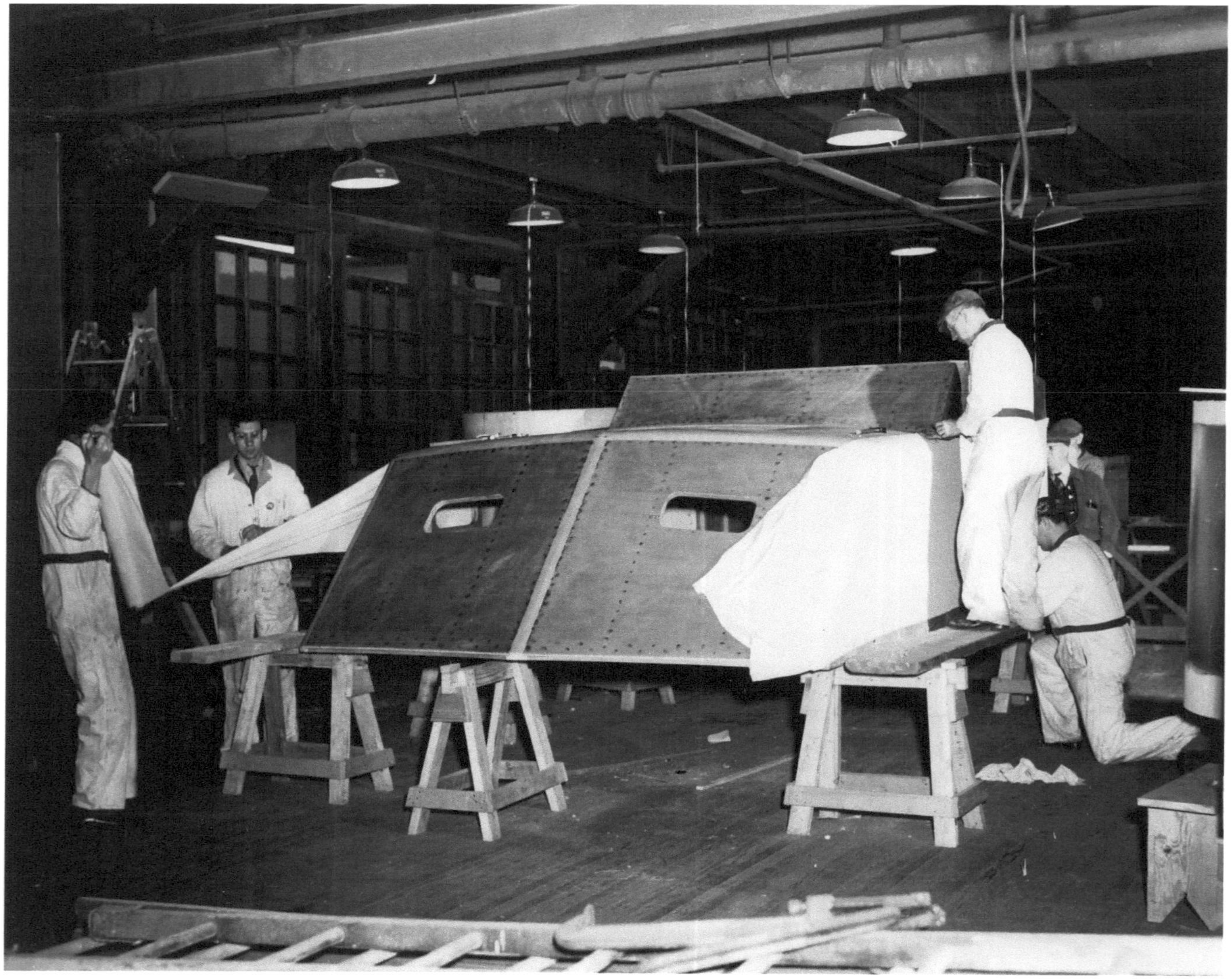

Figure 4-32. *Workers in the main assembly area are working on one of the boats chart house cabin structures. They are stretching air plane cloth over the entire area which will then have glue applied under heat to insure that no leaks occur. (PT Boats Inc.)*

Figure 4-33. *These drills were a common heavy duty type used at Elco. They had large Jacobs chuck's, with the shaft having a cutting or honing head to align the bore. It also had a clutch to prevent injuring the operator using it, when taking too much of a bite. This was a custom tool used for drilling out the openings for the boats three propeller shafts. These Elco workers are installing the Starboard Propeller Shaft through the propeller strut, through the cut-less bearing, into the shaft log, and finally into the engine room. Once the other end of the shaft is inside the engine room, the keyed coupling flange can be installed onto the end of the shaft. This whole procedure takes about 4 people to man-handle the heavy monel prop shaft into the correct position and involves using a lot of lubrication (grease or tallow) and then twisting, pulling, and pushing! Only then can the shaft and coupling alignment process begin. It takes several hours to complete the process for three shafts. (PT Boats Inc.)*

Figure 4-34. *This special wax machine was used for shaft seals (stuffing boxes) using Rope that is impregnated with wax. The water exposure from underneath the boat keeps the seal cool enough that the wax won't melt under friction of the shaft spinning, assuming that the "packing nut" (or gland) on the inside of the shaft isn't tightened too much. The packing nut was tightened to a certain tension on the wax rope to only allow a very slow drip. If the packing nut is too tight, it creates too much friction on the wax rope, then and the wax melts and the rope seal will leak and have to be removed, replaced and then properly re-tightened. (PT Boats Inc.)*

Figure 4-35. *The massive hull is in the stages of being turned over. This innovation made it so much easier for workers to work on the hull from above rather than a cramped position below. Once turned over assembly above deck will begin. These workers were responsible for all boat turning and the hock up to the over head wenches as turning straps needed to be placed in strategic areas. At this stage, the hull weighs in at 12 tons. (PT Boats Inc.)*

Figure 4-36. *Elco worker is shown with a long handled iron. This tool was used on the hull to heat the glue used to impregnate the wood. PT Boats were made waterproof by constructing their hulls with a double layer of mahogany planks, with a layer of glue-impregnated canvas sandwiched between them, essentially creating a watertight seal; this design allowed for a lightweight yet strong hull that could be easily repaired if damaged. This long handled iron made it easier to reach areas of the hull. (Irwin Chase)*

Figure 4-37. *This Women is winding motor armatures. The armature is an integral part of the motor (or generator), as this is the component that carries the alternating current (AC) in an electric machine. The armature conducts AC, even on Direct current (DC) machines using the commutator (which reverses current direction every so often) or due to electronic commutation. Armatures are comprised of a number of components. These parts include the core, the winding, the commutator and the shaft. (Naval War College Archives)*

Figure 4-38. *One of several tool cribs at Elco. This crib carried everything from sanders and routers to electrical screwdrivers to skill saws. Workers would check out the tools they were using for each shift, giving the tool crib staff a bronze tag with their employee number on it. In this way they would know who had what tool and who took it out. (Irwin Chase)*

Figure 4-39. *Hull Department Carpenter in the main assembly plant is making his way down the step ladder to begin his noon time lunch break. Although tools were provided by the company, many had their own tool boxes and tools used in the construction for wooden boats. Notice the sneaker type shoes he is wearing, which provided a non slip bottom. The BP scaffolding system was adjustable which permitted unobstructed working space on any tier when framing, planking, or painting. The scaffolding was made for Elco by the Bergen Point Iron Works of Bayonne, New Jersey. (PT Boats Inc.)*

Figure 4-40. *This women worker is operating what was called a nibbler machine. Nibbling is a cutting process which uses a punch to cut multiple holes in the work piece. These holes overlap, meaning that larger areas of material can be removed by nibbling along the desired profile outline. (PT Boats Inc.)*

Figure 4-41. *The Elco Sheet Metal shop produced the many parts needed to complete the assembly of the boats. In this 1943 photo, this women worker is busy operating a press brake machine. It was used to bend sheet metal or other materials into a desired shape. It consists of a stationary lower die or "bed" and a movable upper die or "punch" that presses the material into the lower die, creating the bend. It was commonly used in sheet metal fabrication, metalworking, and manufacturing to produce a variety of bent metal parts and components. (Irwin Chase)*

Figure 4-42. *The fabric or upholstery shop at Elco, made many items for the boats. Here they would construct seat cushions, bunk cushions, and cushions for the Officers quarters and crews quarters. Before being hired at Elco, many of the women were seamstresses before the war. (Frank J. Andruss Sr.)*

Figure 4-43. *With men going off to war, Elco took up the slack with hiring many Women workers. Here these workers are in training, and learning how to correctly hammer, rivet, sand, and work on the boats hulls and planking. (Irwin Chase)*

Figure 4-44. *This worker in Building 21 is using a bandsaw. That skinny blade makes it easier to cut curves and details. It had a long, flexible blade made of toothed metal that was used to cut materials. They were versatile and could perform a variety of tasks, such as ripping lumber, cutting curves, and making veneer. (Irwin Chase)*

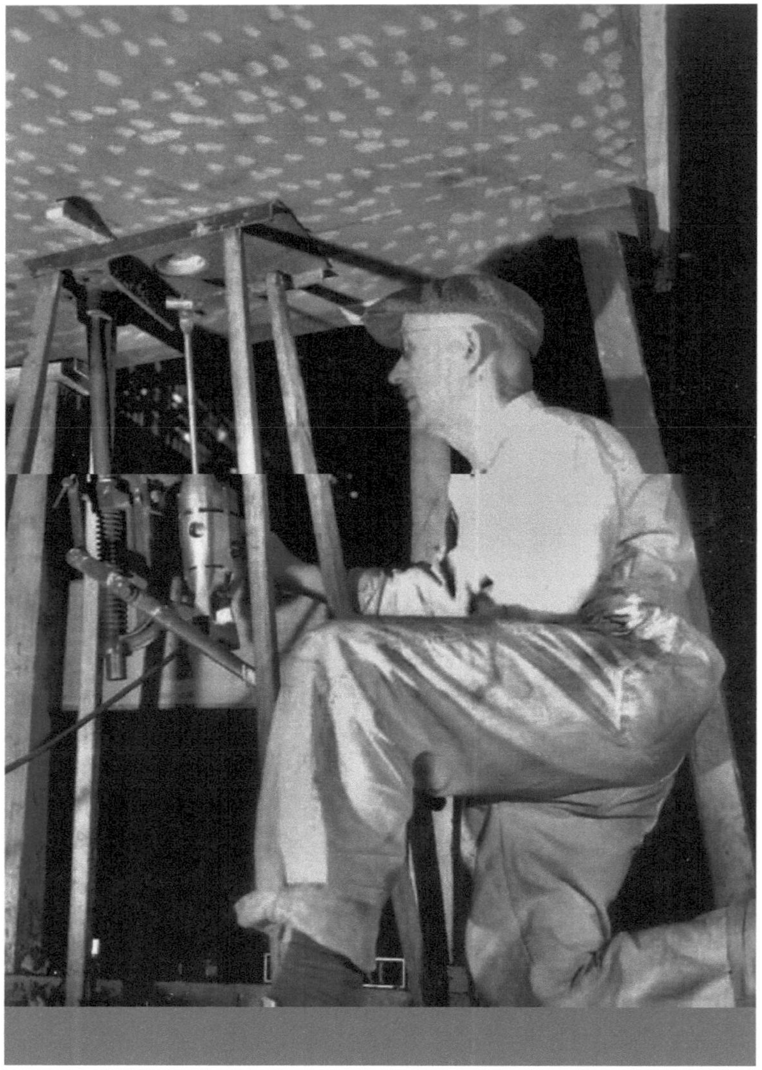

Figure 4-45. *Another part of the assembly process was the drilling and placement for the boats three bronze struts. The struts are a key part of a boat's design, supporting the propellers and affecting how water hits them. This can impact the boat's performance and noise levels. The strongest struts are usually made of cast silicone bronze or commercial bronze. Using a power drill and template, placement locations had to be 100% accurate. (Irwin Chase)*

Figure 4-46. *PT-375 has been towed into the wet boat basin. Here weapons systems, ammo boxes, torpedo racks, and other sub assembly took place. In this photo clearances are being checked for the starboard side 50 caliber machine gun mount, as well as the view from the helm area looking forward. (PT Boats Inc.)*

Figure 4-47. *Elco's oldest employee, Frederick Bulkeley (right) was hired on April 16, 1942. When Bulkeley tried to enlist in the Navy, he was told that he was too old. Newspaper's in the Bayonne area carried this story and shortly after, Bulkeley received over 200 offers for employment. At 76 years of age Bulkeley went to work doing his inspection job, as well as promotional work for the company. Here he inspects one of the boats helm wheels. Frederick's son was the famous Lt. John D. Bulkeley who was awarded the Medal of Honor for his work with MTB Squadron 3 in the Philippines, early in the war. (Charles B. Jones)*

Figure 4-48. *Foremen and workers in the sheet metal department are being trained in the art of welding techniques. (Irwin Chase)*

Figure 4-49. *Workers are using a pleasant day outdoors to Rivet the large girders used on the boats. These girders provided maximum strength and stability in all types of weather the boats might encounter. (Irwin Chase)*

Figure 4-50. *This worker is running a Baxter Whitney table spindle molding machine. It had two blades that clamped into the head, and would cut different shapes or slots into the timber, these machines can run up to, or even over 2,2000 RPM. (Irwin Chase)*

Figure 4-51. *Employee checking out one of the Goodrich Cutless Bearing containers. These were a water lubricated, rubber-based shaft bearing. The bearings were used at the aft end of the boats shaft tube, to support the propeller shaft. (Charles B. Jones)*

Figure 4-52. *This employee is using a hand powered drill with a long shaft bit to drill the hole for the eye towing and installation gear. This installation was completed on the bow of the boat. The eye towing gear was essential should the boat need to be towed after damage in battle, or upon engine failure.(Irwin Chase)*

Figure 4-53. *Scandinavian worker Olav Nilson. Olav worked at Elco for many years and worked on the keels of the PT boats, using an adz to shape and cut the keel of each PT boat. Elco had hired one other Scandinavian worker that also worked on the boats keels. Olav stayed with the company until it closed in 1948. (PT Boats Inc.)*

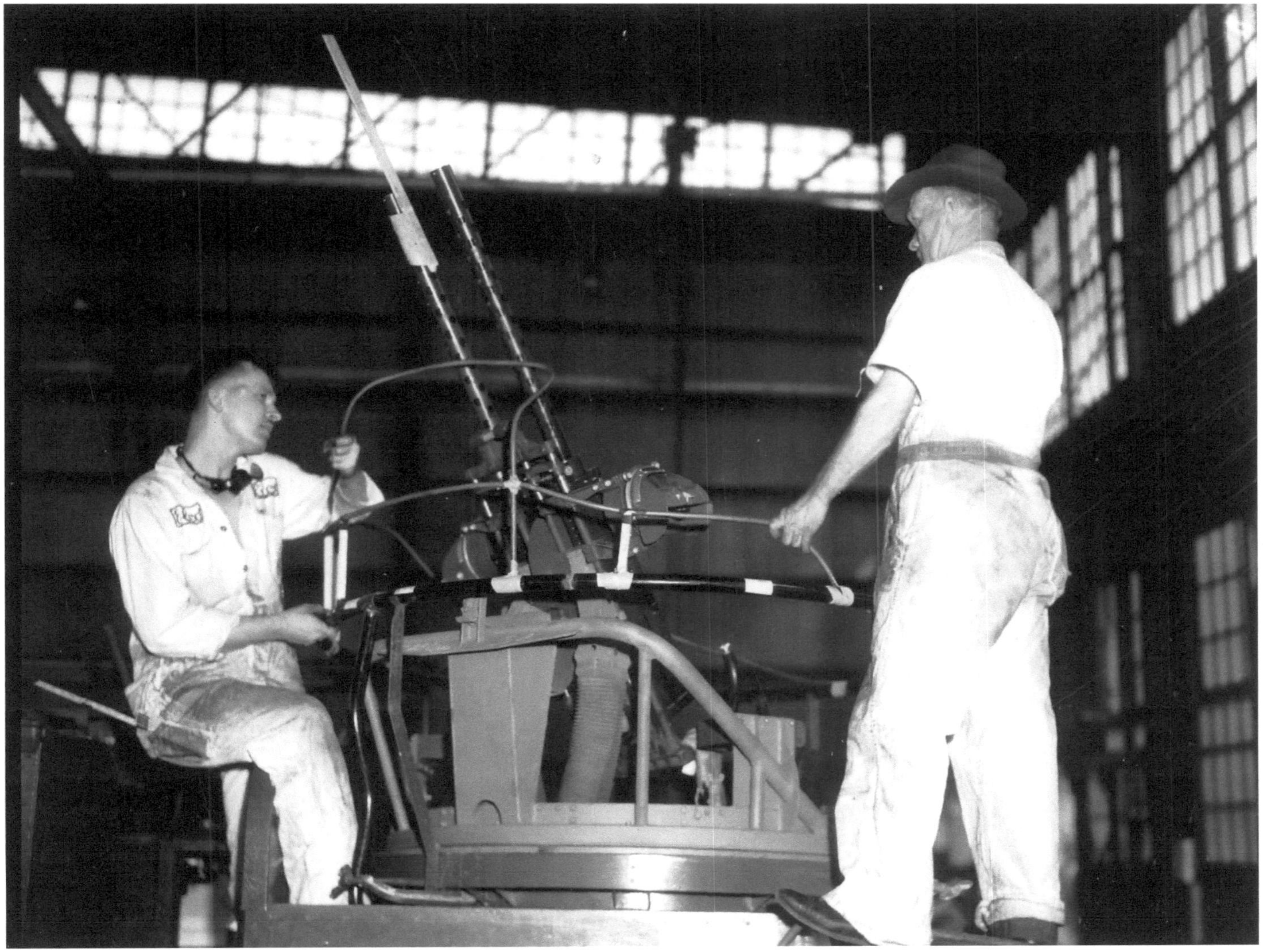

Figure 4-54. *Workers are making measurements and adjustments to the starboard side gun turret depressor rails. Machine gun limiting stops of brass or steel tubing was fitted as required to prevent accidental firing into the boat or occupants on the bridge. The 0".50 Mount, Mark 17 Mod. 1 was a hand-operated machine-gun mount, carrying two caliber 0.50 Browning aircraft machine guns. Each boat carried two mounts starboard and port side. Two canvas covers were provided for .50 cal. machine gun mounts to completely enclose guns, mounts, and limiting stops. After turret had a separate spray shield to the height of limiting stop. All canvas covers and life preservers received the approved treatment for fire retardant. (PT Boats Inc.)*

Figure 4-55. *This worker from the Paint Shop is spraying the hull of the boat. The hull was first primed and finished with three coats formula 511 special haze gray, with the final coat in accordance with latest Elco instructions. The bottom was also primed with copper bottom enamel, fastening holes cemented with surfacing putty and finished with two coats copper bottom enamel. (PT Boats Inc.)*

Figure 4-56. *Women workers from the Electrical Assembly Department are busy wiring electrical boards. The first two women on the first bench closets to the camera are installing RPM and Manifold Pressure gauges to the boats main helm control panel. (PT Boats Inc.)*

Figure 4-57. *This worker is under the hull adjusting one of the three struts for the boats shafts and propellers. Struts played a very important part in the design of the boat. They support the propellers, and also affect the way the water hits the propeller, which can have a significant impact on performance and noise. These struts were made of unpainted manganese bronze with three fitted to each boat. (PT Boats Inc.)*

Figure 4-58. *A rare look at the backside of the Elco plant, which had a rail line coming into the property. Daily freight cars would be unloaded here at this platform for many dropping off hundreds of supplies needed for the boats. Smoke generators would be filled here at this location and then brought out to the assembly area for placement on the boats. (PT Boats Inc.)*

Figure 4-59. *Elco would build their own Cafeteria for workers, making it more efficient for three shifts of workers to get a sit down meal. Access to a good meal at the factory cafeteria was crucial for maintaining worker morale, especially with long hours and demanding work conditions. Although no evidence was found at Elco, "Victory Gardens" were sometimes planted right on the factory grounds to help combat food rationing. These gardens were specifically intended to provide fresh vegetables. Here some of those workers and cooks have made a cake for one of their own.*

Figure 4-60. *With the development of the Elco Turret (Thunderbolt), they would construct a new building in 1943. This would become Building 29 where these turrets would be assembled. The turret housed four 20mm cannons. It was a single-person electric powered turret that went through several iterations. These turrets boasted a remarkable capability where a single operator could unleash a formidable barrage of bullets towards approaching enemy targets. (PT Boats Inc.)*

Figure 4-61. *Inside building 29, workers on the assembly line are working on the carriage mounts which will house the 20mm cannons. To the right a shop foreman is inspecting the work, while other assembly, like adding the operators seat is being completed. This photo is from the restart of the Mark 15 Mod 1 Elco Turret production line in 1945. Elco had delivered 26 of these updated mounts to the Navy. These were being produced for the Invasion of Japan combatants. Production had been halted (suspended date was December 4, 1943) for approximately 8 months in 1944 in order to evaluate the design and provide changes. Updates continued during the restart. (PT Boats Inc.)*

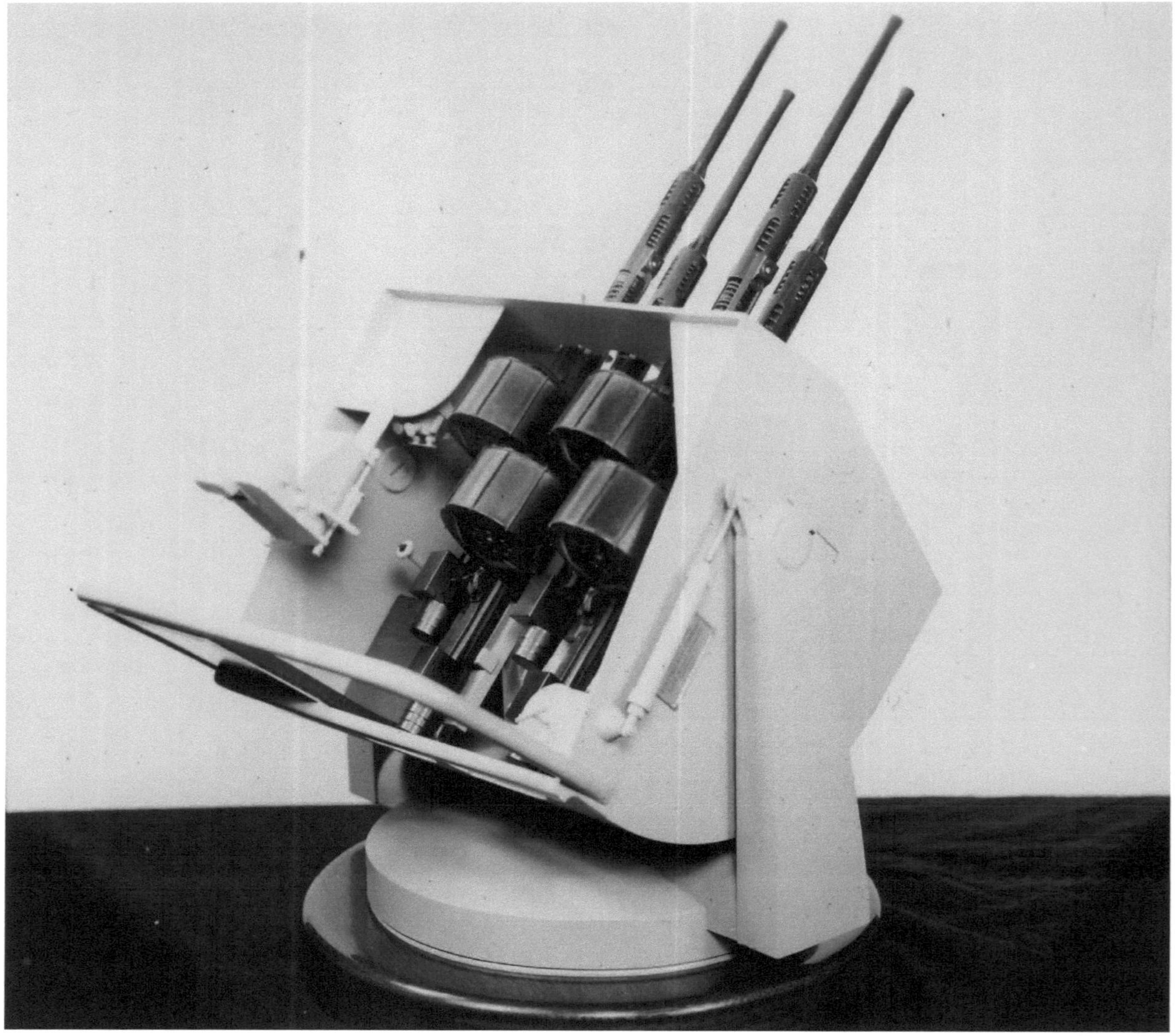

Figure 4-62. *A look at the Elco turret (Thunderbolt) system showing the four 20mm cannons. Yet to be added are the gun site and operators seat. Other components will be added to complete the turret, but the reader will get a true sense of what this system looked like before being mounted on a few of the boats. Original order from the Bureau of Ordnance called for 200 of these mounts to be constructed, but on May 5, 1945 this was downgraded to 120. (PT Boats Inc.)*

Figure 4-63. *After touring the boats chart house and crews quarters, Ensign Eliabeth Gardner, USNR makes her way topside. Gardner was in the administrative support and PR with the Navy. This is one of the newly built 500 class Elco PT boats, sitting in the boat basin in front of Building 21. (Irwin Chase)*

Figure 4-64. *In August of 1945, Elco was winding down from the war. They had several hulls in different phases of construction that were canceled. They had to do something with these hulls, and set about cutting them up. Here in Building 21 workers are sawing the hulls in sections for removal. (Charles B. Jones.)*

Figure 4-65. *With the hulls cut up into sections, they were trucked outside of the plant and then hauled to an undisclosed location for burning. Here we see the hulls which have been placed onto small carts on wheels, where they await removal from the plant. (Charles B. Jones.)*

Chapter 5

The Lighter Side of War

With any company the product that gets produced, can be measured by the people that produce it. Elco was no different and during the war years, those workers had an incredible appetite for patriotism.

They were building the little PT Boats and were doing their very best to rush these little Midget Sluggers to the war zone. Elco had a long standing tradition in wooden boat building, and many of the pre-war workers were obtained to continue their excellent work. It was felt that Elco produced the greatest of the wooden boats before the war, and their reputation followed them into World War II.

With so many workers running three shifts, it was felt they needed something more to take their minds off of the rugged task of building boats. Elco's Labor Management would have the task of providing an escape of the every day rigors of their jobs.

The Labor Management committee was formed at Elco around April of 1942, it's efforts promoting maximum production of the PT boats, and at the same time, endeavoring to combine safe, healthful, and congenial working conditions. The committee combined four Labor representatives and four Management representatives. Their combined efforts through the war years brought about splendid teamwork.

First they added suggestion boxes and placed them in all locker rooms, and shop departments, for all workers to submit ideas in monthly competition to improve production, save materials, conservation of tools, and equipment. Workers enjoyed this and Management awarded prizes of First prize, twenty-five-dollar War Bond; Second prize ten-dollar War Bond; Third prize five-dollar War Stamps, with duplication of awards if necessary. A certificate of award accompanied each prize. Suggestions also had a chance to be submitted to the War Production Board at Washington and receive national recognition with "Meritorious Awarc" certificates which were handed to the winners.

Elco was awarded the Army and Navy "E" Flag for high efficiency through cooperation from all of the departments. They would received this award six times during the war years.

Another morale booster for the employees was the War Service fund Committee. They put out the word about fun packages with money collected, directly from the Elco employees. They were able to pack and ship razor blades, candy, writing paper, shaving cream, and special Army/Navy pencils. These items went directly to the Elco boat squadrons in the war zone.

Another fine program was the Apprenticeship program, which would be awarded to many Elco employees since its inception in May of 1941. This program was done to help with plant production. They would embark on a two year schedule of training, the school work, and study accomplished at the Vocational High School in Bayonne, and the actual mechanics work put into practice at the plant. Many of the well trained men became full fledged mechanics serving in the many shops at the plant.

Transportation problems for the many Elco employees were many and varied, particularly those who relied on gasoline for their motive power. The drastic curtailment of gasoline had been at times critical for the war effort. Thanks to Elco management, gasoline was always available to Elco employees, to the extent of their OPA award, at the gas station assigned exclusively for Electric Boat Company War workers. It was a unique arrangement, coupled up with car pooling and riding, and assisting employees with their applications for gasoline and tires.

During the summer months, many successful programs were arranged, with lively noonday crowds of employees, who would gather around the outdoor band-stand on such occasions as Flag Day, Fourth of July, War Bond rallies, Suggestion rallies, and Army and Navy E Awards. These rallies included many prominent speakers, Navy and Army officers, and heroes; Metropolitan Opera Stars, and particular emphasis on Commander John D. Bulkeley, USN. Many programs included plant talent such as the singing Elcoettes, three lovely Elco employees, and Labor Management speakers and presentations of certificates of award.

Elco also presented Noontime Educational Films. These films with sound provided yet another way for employees to relax. These films showed the world wide events of the times and showed the daring deeds of our boys at the front. Labor Management also had on display in Washington D.C., in February 1944, two outstanding posters called "The Wheel of Success" and The Pinnacle of Production", which attracted nationwide publicity for Elco.

Monday musicals were another way for employees to relax and enjoy vocal talent from Elco employees. These weekly events were done on the outdoor band stand, weather permitting. Many of the vocal talent were trained under the supervision of Messers, P. Harding, and E McCurdy. In addition to the vocal talent their were also an array of musicians who shared in these programs.

Elco employees also had what was known as the Red Cross Drive, which raised considerable money for the Bayonne Chapter of the American National Red Cross. Money raised was given to the fighting allies.

There was great enthusiasm at Elco when Labor Management formed Sports teams. One of the very popular ways to relax and have fun was playing baseball. A field was constructed, adjoining Building 21 (main assembly) for use during noontime and after working hours. Elco also formed softball teams playing in the Industrial league. Elco would win their first championship in 1941. This would be Elco's first championship for the men's softball team. Elco would win the championship again in 1942 and 1944. Trophies and sweaters were handed out to the winners.

Elco would also form a basketball team, winning the 1941–42 Championship. They played in the Hudson County Amateur Athletic League, and gathered the top spot in the payoffs by beating General Cable Company in the finals. Early basketball courts were rarely perfectly smooth. Knee pads were standard equipment to prevent severe injuries from protruding nails, splinters, uneven planking, and other potentially dangerous defects in court flooring, which were even played outdoors.

Probably the highest turnout of employees was for the Inter Departmental League Bowling teams. Many teams formed were from different departments, consisting of teams from the Plumbing, Engine, Sheet Metal, Machine, Control, Boat Builders, Riggers, Joiners, Electrical, and Store Room Departments. Thirty-two teams comprising of some 160 players, would play their games at the Playdrome, at Union City. They played at over thirty-two bowling alleys. Women also made up several of the teams.

Elco also played benefit games to support the War Service Fund. One basketball game in particular was played at the Bayonne Junior High School between some of the Long Island hometown boys, and Elco's 1944 Industrial Champions. Over four hundred tickets were sold for the event, with Elco winning 60–38. They raised $176.00 for the Service Fund.

Later in 1945, Golf would become another passtime for employees. Although the supply of golf clubs was curtailed for the war effort, there was roughly thirty employees out on the links, during some of the outings.

Elco also set up many parties, complete with dancing. Elco workers would attend barn dances, held at Mackenzie Post on Avenue E Bayonne, New Jersey. Guests would listen to country music and even square dance to live music of the day. Barn dances and social events like this, helped boost the morale of workers. These dances weren't just for fun—they helped workers cope with the emotional toll of war and reminded them of the strength of community. They also attended engagement, Halloween, and Christmas parties as well.

Elco workers enjoyed their annual outing as well, where they would board buses, heading to fun places. Outdoor games were played, complete with hot dogs, hamburgs, and other good food. Later they would head to the Pavonia Yacht Club in Bayonne, New Jersey, to enjoy a dinner, and play golf.

War Bond rallies were an important part of the makeup of the Elco Naval Division. Bond drives in many states to help raise money to build PT boats was a common practice. When they did raise enough funds, a ceremony at the plant took place and that boat was put on display.

One of the highlights for the company was the awarding of the fourth

Army Navy award for excellence and the World premiere of the short film "Devil Boats". This ceremony and showing was held on August 22, 1944, at the Dewitt Theater, Bayonne, New Jersey. Over twenty-eight hundred Elco employees and invited guests attended this special night. The American Legion Band of Richmond County, provided a concert as guest's were arriving. This short film was created by Warner Brothers with Elco's express permission. It remained a highlight for those employees that toiled long hours building these boats, something they were very proud of.

The "Salvage For Victory" campaign asked that all Americans do their part in supporting the new war effort by collecting, saving, conserving, and recycling materials that could be repurposed for military uses. In Bayonne and New York, mayors were asked to save the same kinds of materials from municipal dumps and incinerators. In Bayonne, the Department of Sanitation began picking up materials on an alternating schedule with regular garbage collection. Elco had workers from the plant in being the nights special Bar Tenders to help raise money and awareness of the program.

One special lunch time entertainment was honoring Commander John D. Bulkeley, USN. He addressed the employees and revealed how well the PT boats were doing in the war zone, and that their boats and hard work was paying off. Frederick Bulkeley, who was an inspector at the plant and the Commander's father participated in the excercises, during which Commander Bulkeley was presented a model of his famous PT-41, in which he rode in the Philippines. This beautiful model was made by Elco employee Don Rosencrantz. PT-41 was the boat in which General MacArthur was taken from the Philippines, before the invading Japanese forces overtook the island.

It was by 1945 that Elco would have their seventh War Bond Drive, collecting some fourteen million dollars. Due to past performances this drive was featured by rallies, posters, publicity, and speeches. Thousands of dollars were set aside each pay day through payroll savings. Co-operation of activities by Labor and Management, helped Elco reach their goals over the years, for the war effort.

The Elco Works Employees Association was incorporated under the State laws of New Jersey, on July 1, 1937, as a Union for collective bargaining purpose with the Elco Naval Division of the Electric Boat Company. During the war years, the association brought to the employees, a strong union, with a clause in 1943, covering the men who entered the Armed Forces. This was prior to the GI Bill of Rights.

One interesting form of fun for the employees was Horse Shoe Pitching. It was set up below the Elco cafeteria, close to the south railroad fence. Many employees used their lunch hour to get some fresh air and play this fun game. It was decided that the two best employees playing this game were Wilbur Bowman, and Jim Dunlop.

More fun was added when a ten cent dance was held at the American Legion in Bayonne. For ten cents you could get a dance with a pretty lady, and in this case the girls were workers at the plant. The money raised for this type of dance were placed into the war relief fund or to raise money for war bonds. Also a mock jail was set up, and you had to get raised funds to get out of Jail.

The Labor Management Committee did an incredible job of providing vast escapes from the everyday life, of building boats. Employees could do many recreational forms of enjoyment, and it was because of this that Elco had employees that worked hard to produce Uncle Sam's PT boats. It was President Roosevelt who pointed out that Human Beings could not sustain, continued hard work, for very long without proper balance between work on one hand, and vacation and recreation on the other. Elco workers, both men and women for whom adequate recreation provision was made, returned in better condition to meet their tasks. Elco knew that their workers were pretty much all alike, in their relish of the golden moments for relaxing play, when that time card was punched and the work shift ended.

Figure 5-1. *Elco receives it's first Army Navy E Award for product excellence. Officers and other dignitaries attend this special event. Holding the flag to the left is Elco's Executive Vice President Henry R. Sutphen, to his left is Assistant Supervisor of Ship Building Lt. Frank W. Rowe, to his left is Elco employee, Milton W Hoyt, an employee of 38 years, and behind him is Preston L. Sutphen, General Manager of Elco. This award would be presented to Elco six times during the companies assembly of PT boats during World War II. (Lt. Cmdr. F.W. Rowe)*

Figure 5-2. *Hundreds of workers, office staff and guest's have crowded Building 21, the main assembly building at Elco. They are here to attend the ceremonies for the first presentation of the Army Navy E Award for excellence. (Lt. Cndr. F.W. Rowe)*

Figure 5-3. *An original Army Navy E Awarded indoor flag, presented to the Elco Naval Division for it's 5th award. This one belonged to Managing Constructor Irwin Chase and hung in his offices at the plant. It belongs to the author and was donated by the Tremaine family. (Frank J Andruss Sr.)*

Figure 5-4. *This Elco Award booklet was handed out to those attending the companies award of yet another Army Navy E Award. It is dated August 26,*

Figure 5-5. *Elco workers from the Production Department are enjoying a night out, as part of the ceremonies for the Army Navy E Award. They are at the Tise Tavern in Jersey City, a popular spot and frequented by Elco for dances and dinner parties. (PT Boats Inc.)*

Figure 5-6. *A look at the 1941 Industrial Softball Champions. This would be Elco's first championship for the mens softball team. They were led by their ace Vince Delikat that pitched them to victory in the playoffs. Members that are known include Mike Banasiak last row extream right, Joe Pruskiewicz in the middle row, second from left, and Walter Cambell holding the sign. Elco would win the championship again in 1942 and*

Figure 5-7. *Beautiful trophy for the 1941 Industrial Softball Champions. Presented to the team by the Elco Works Employees Association, the trophy would be placed inside the lobby area of Building 20, which was the main offices of Elco. (Irwin Chase)*

Figure 5-8. *The 1942 and 1944 Industrial Softball Champions now lined up and looking sharp in their Elco presented Championship sweaters. Third from the right is Charles Waller, executive assistant to the Works Manager. The women in the center is Miss Mary Patterson who worked in the offices in the Works Department. (Charles Waller)*

Figure 5-9. *Championship Sweater presented to team player Charles Waller, as the 1944 Industrial Softball Champions. Waller lead the team that year in hits and runs batted in. This sweater was donated to the Author by his son James Waller, and is a wonderful example from many years ago. (Charles Waller)*

Figure 5-10. *A nice photo of the Inter Departmental League Bowling Champions for the 1942-1943 season, showing the men's team from the Electrical Department. A large trophy was presented to the captain of the team, from Mechanical engineer George A. Fulton, 3rd from left. (PT Boats Inc.)*

Figure 5-11. *Winners for Bowling team from the Electrical Department line up behind the transom of this boat, just before the boat is turned over. With trophy in hand, the sign reads "Lets Bowl Em Over, whether in the workshop or on the ally's, Elco men strive for Victory, if we keep the ball of endeavor rolling straight, we are bound to scatter the pins of indifference, Men and Women of Elco, let's Bowl Um Over". Shaking hands on the left is George A. Fulton, one of Elco's mechanical Superintendents. (PT Boats Inc.)*

Figure 5-12. *Members of the 1941-42 Elco Championship Basketball Team. They played in the Hudson County Amateur Athletic League, and gathered the top spot in the payoffs by beating General Cable Company in the finals. Early basketball courts were rarely perfectly smooth. Knee pads were standard equipment to prevent severe injuries from protruding nails, splinters, uneven planking, and other potentially dangerous defects in court flooring, which were even played outdoors. (PT Boats Inc.)*

Figure 5-13. *Members of the Bayonne Industrial League Elco Basketball Champions for the 1943-1944 season, take time to pose next to Building 21. Front row: Nazurak, LeFurge, Gendzwill, and McCarthy. Back row: Goodhart, Orlowski, Britt, Delikat, Grimm and Eggers. (Frank Karahuta)*

Figure 5-14. *Great look at the Elco Band. Made up from workers in different departments, these nine players certainly would provide wonderful performances, during the noon time Labor Management functions. Leading the band on the accordion is J. Ckstowicz. Notice all the players still wearing their company provided work uniforms. (PT Boats Inc.)*

Figure 5-15. *Ladies from the inner offices at Elco, surprise Secretary Mary Morgan (3rd from left) who was recently engaged to be married. They have given her a cake and for a wedding present a wonderful set of silverware. Workers took the time, usually at lunch break to have these small gatherings for one another. (PT Boats Inc.)*

Figure 5-16. *This beautiful cake was presented during another noon time ceremony in the offices of Building 20. The cake says "Good luck to our Ensign from all of us". The boat pictured on the cake is PT-4, which is a bit odd as the boat was built by the Fisher Boat Works, Detroit, MI. To the far left is Charles Waller, executive assistant to the Works Manager, who at that time was Glenville Tremaine. (Charles Waller)*

Figure 5-17. *A noon time War Bond rally being held at the outdoor stage. Addressing the workers is Lt. Cmdr. F. W. Rowe, the Assistant Supervisor of shipbuilding at the Elco Naval Division. The band has also taken the stage, as always providing good music for those in attendance. The outdoor stage was generally set up behind Building 16, the wet boat basin. (Lt. Cmdr. F. W. Rowe)*

Figure 5-18. *Special office inside Elco for Labor Management. As one can see it has many war time posters and War Bond posters. To ensure uninterrupted production, Elco workers generally agreed to a "no-strike" pledge for the duration of the war, in exchange for government intervention to resolve labor disputes fairly. Labor Management worked hand and hand with workers to ensure happy workers, and help with any disputes. (Navy War College Archives)*

Figure 5-19. *Elco workers attend a barn dance at Mackenzie Post on Avenue E Bayonne, New Jersey. Guests would listen to country music and even square dance to live music of the day. Barn dances and social events like this, helped boost the morale of workers. These dances weren't just for fun—they helped workers cope with the emotional toll of war and reminded them of the strength of community. (PT Boats Inc.)*

Figure 5-20. *Elco's Labor Management Team set up suggestion boxes throughout the plant for production ideas. Prizes were awarded to those with the best ideas. First place received a twenty-five-dollar war bond. Elco over the years would receive some very good suggestions for operating the plant, and inproving Departments. (Irwin Chase)*

Figure 5-21. *An engagement party being held at Mackenzie Post on Avenue E, Bayonne, New Jersey. Helen Michael's second from left is, all smiles at the party thrown for her, from the workers of the Production Department. The women 4th from Left second row is Alice Messeder, also from the Production Department. (PT Boats Inc.)*

Figure 5-22. *Eleanor Behney (ready to turn the hull) took a job in the design business office of Elco. While there Eleanor was first honored by Electric Boat, selected to be the ship's sponsor of PT-138 and preside over the "turning of the hull" ceremony, a major milestone in construction of PT boats. After the war, she was sought after by the company as a model and featured in several advertising photos and brochures promoting Elco's famous luxury yachts. (Irwin Chase) <div> </div>*

Figure 5-23. *Wooden sign made by the carpenter shop for Elco employees in their attempt to purchase two PT boats. Sponsored by the Labor Management War Bond Committee. Elco workers did buy enough war bonds to make the purchase of the boats, showing their patriotism and determination. (PT Boats Inc.)*

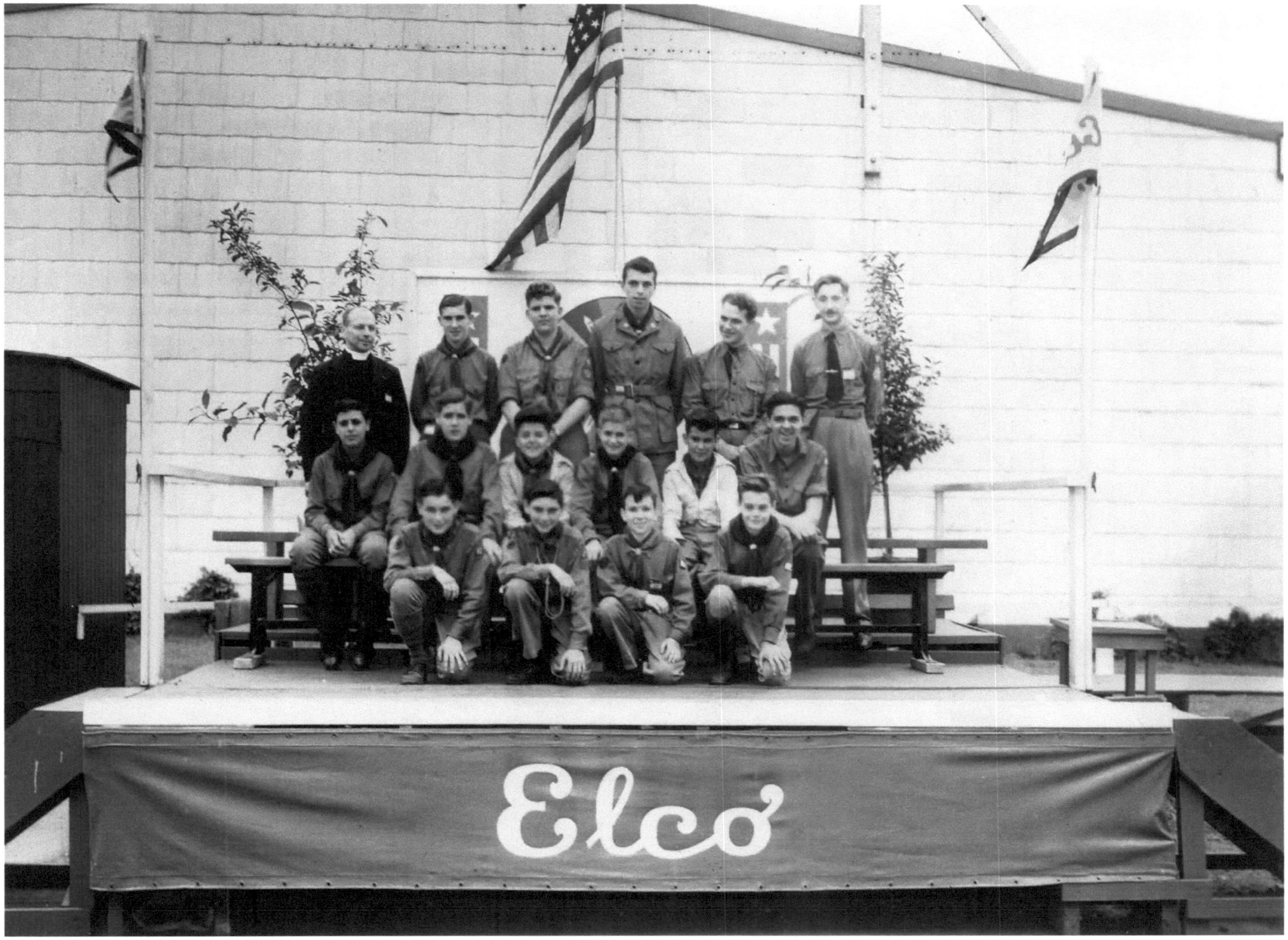

Figure 5-24. *Boy Scout Troop 25, takes the time to visit the Elco Naval Division, where they were given a tour of the assembly areas. On July 15, 1941, the scout troops of Bayonne organized an emergency corp of 800 scouts to aid in the civilian defense of the city. Troop 25 was sponsored by St. Vincents Catholic Church loated at 979 Avenue C, Bayonne, New Jersey. (PT Boats Inc.)*

Figure 5-25. *Labor Management noon time entertainment on the outdoor band stand. From left: Works Manager Glenville Tremaine, Comdr. John D. Bulkeley of squadron 3 fame in the Philippines, and Managing Constructor Irwin Chase. They have presented a model of Bulkeley's PT-41 as part of the entertainment program. This beautiful model was built by Elco Hull Department employee, Don Rosencrantz. (Irwin Chase)*

Figure 5-26. *Executive Vice President of Elco, Henry R. Sutphen has taken to the stage, which was set up inside Building 21. He is addressing workers as to the fine job that they are doing in producing our PT boats for the US Navy. Also, he is addressing them as to the War Bond Drives that were producing great results. Behind him in the first row (second from left) is Elco's Managing Constructor, Irwin Chase. To Chase's left is Lt. Cmdr. John D. Bulkeley, from PT boat Squadron 3 fame, and to Bulkeley's left is Elco's General Manager, Preston Sutphen. (PT Boats Inc.)*

Figure 5-27. *Elco devised an ingenious way to honor each state of the union, by naming a boat after that state. Here men and women workers pose behind the Massachusetts sponsored boat, PT-358. In a letter dated May 13, 1943 the Governor of Massachusetts, Leverett Saltonstall thanked Elco, and was proud that this fighting boat would represent his State. (PT Boats Inc.)*

Figure 5-28. *The Army Navy E Flag proudly flies over the Elco Naval Division of the Electric Boat Company each and every day. Here Police attach the flag ready to hoist it up. You can see the five stars for excellence in productivity that the company was awarded during the war years. (Frank J Andruss Sr.)*

Figure 5-29. *Elco workers enjoy their 3rd annual outing from 1943. From the look of things they are having a great time. Later they would head to the Pavonia Yacht Club in Bayonne, New Jersey. Committee chairman, Edward McCaffrey, Vice President of the association reported a stellar turnout. (PT Boats Inc.)*

Figure 5-30. *Another game of softball during the noon time break at the plant. From the looks of this photo the man at the plate is swinging for a home run, while others watch. (PT Boats Inc.)*

Figure 5-31. *Noontime baseball games on the Elco field became very popular. After hours games sometimes lasted into the night. These games managed to draw pretty good crowds as well. Here it looks like this runner is going to be called out at the plate. (PT Boats Inc.)*

Figure 5-32. *These beautiful singers called the Elcoettes, take to the stage for the ever popular noon time entertainment, this time featuring popular songs of the day. These three ladies were also employees of the plant, worked in the inner offices. Elco workers dressed in their white company supplied coveralls sit up on the outdoor stage, as company officials will address the workers, about war bonds and their commitment in producing the best boats for the Navy. (PT Boats Inc.)*

Figure 5-33. *Halloween time in 1943 has these male workers dressed as ladies. Even more amazing is that they are showing their legs on the outdoor stage, during a noontime break for workers. All in fun, it helped break the tensions and rigors of assembly work, and gave the workers time to relax and clear their head. (PT Boats Inc.)*

Figure 5-34. *War Bond rallies were an important part of the makeup of the Elco Naval Division. Bond drives in many states to help raise money to build PT boats was a common practice. When they did raise enough funds, a ceremony at the plant took place and that boat was put on display. Here representatives proudly display this boat which was sponsored and purchased with War Bonds by the people of the City of Rochester and Olmsted County of Minnesota. To the left is Lt. Cmdr. Frank W. Rowe, the Assistant Supervisor of Ship Building at Elco. (PT Boats Inc.)*

Figure 5-35. *More representatives and sponsor officials crowd the hull of this boat, during the War Bond Hull Ceremony from the City of Rochester. In the front row far right is Elco's Managing Constructor Irwin Chase. Behind Chase is General Manager of Elco Preston L. Sutphen, to his right is Executive Vice President of Elco, Henry R. Sutphen, and to his right is Elco's Assistant Supervisor of Elco, Frank W. Rowe. (PT Boats Inc.)*

Figure 5-36. *War Bond representatives during ceremonies, were allowed to tour the boats so they could see what their purchase has built. Here one of the representatives checks out the helm area of this boat while it is docked. (PT Boats Inc.)*

Figure 5-37. *These lovely ladies are representatives from the War Bond and Stamp Committee, representing the City of Grand Haven, Michigan. Their city has raised enough War Bond funds to build one Elco PT boat. They are on the hull of PT-375, the boat that was built with these funds. The boat is docked at the outdoor boat shed in front of Building 21. (PT Boats Inc.)*

Figure 5-38. *A wonderful look at PT-375 loaded with sponsor's and representatives from Grand Haven, Michigan. As part of the ceremonies the boat will be taken out for a run just outside of the plant, by the Elco trial crew. Here they will get to see first hand what the boat they raised funds for can do. The boat is leaving the dock from Building 21, the main assembly building. Notice the large twin boom crane over looking the boat basin. (PT Boats Inc.)*

Figure 5-39. *Standing on the bow of the newly built PT-129. Left is Executive Vice President Henry R. Sutphen, Admiral Royal Eason Ingersoll, Commander-in-Chief, Atlantic Fleet, and F.W. Rowe Assistant Supervisor of Ship Building for Elco. (Irwin Chase)*

Figure 5-40. *This officer is attaching a plaque inside the boat which reads, This Fighting Ship sponsored and made possible by War Bond Purchases of the people of Rochester, Minnesota. Each boat would have one of these type plaques attached to her, if purchased by a City, whose people raised enough funds to do so. (PT Boats Inc.)*

Figure 5-41. *A huge turnout along the concrete roadway, as Elco celebrates the launching of PT-379, as she is draped with flags from bow to stern. Other boats in the basin include PT-383 in the center and PT-378 closest to the gasoline barge. The photo was taken on August 3, 1943. These boats in the basin would make up part of Squadron 28, going on to serve in the South Pacific and Southwest Pacific areas. (PT Boats Inc.)*

Figure 5-42. *General Manager of Elco, Preston L. Sutphen, presents a War Bond Trophy to Edward Shibley. Contests were held, and prizes were awarded to individuals with the highest sales. Also on the outdoor stage we see a man and women wearing crowns that say Elco across the front. These were awarded for the second most War Bond sales, along with a certificate of appreciation. By 1945, employees would purchase some $1,500,000 worth of War Bonds. (PT Boats Inc.)*

Figure 5-43. *The Elco Memorial was set up between the main offices. It was in honor of those that worked for the company, but left to join the military branches of the service. Several during the war lost their life, and Elco produced a Magazine called "Those Who Serve". Inside were the photo's of past Elco employees that gave their life for freedom. (PT Boats Inc.)*

Figure 5-44. *Lt.(jg) Joseph Atkinson, boat captain of PT-586, attaches a bronze plaque on the inside of the boats day room cabin. This boat was made possible by Carter County Tennesse War Bond purchases. Helping him with this task is Mrs. Alex Shell of the women's division of the finance committee. Ceremonies followed and the party was entertained at dinner by officials of the company and officers of the Navy. Representing the Elco company was Henry R. Sutphen, Executive Vice President; Preston L. Sutphen and S.C. Kelly, Vice Presidents and Irwin Chase, Managing Constructor. (Frank J. Andruss Sr.)*

Figure 5-45. *This is the original plaque that was mounted on PT-586 by boat captain Lt.(jg) Joseph Atkinson and Mrs. Alex Shell from War Bond Purchases made by Carter County Tennessee. It is from the personel collection of the author and remains a wonderful piece of Elco history. (Frank J. Andruss Sr.)*

Figure 5-46. *One of the highlights for the company was the awarding of the 4th Army Navy award for excellence and the World premiere of the short film "Devil Boats". This ceremony and showing was held on August 22, 1944, at the Dewitt Bayonne, New Jersey. Over 2,800 Elco employees and invited guests attended this special night. To the left of the photo is the American Legion Band of Richmond County, which provided a concert as guest's were arriving. This short film was created by Warner Brothers with Elco's express permission. (Frank J. Andruss Sr.)*

Figure 5-47. *Dignitaries and Navy brass begin to arrive to the Dewitt Theater. Getting out of the Elco car (adjusting his jacket) is Elco's Executive Vice President Henry R. Sutphen and to his left in the hat is his wife, Susanna Preston Lees. (PT Boats Inc.)*

Figure 5-48. *Taking time for the camera while inside the Theater is Left: General Manager of Elco, Preston L. Sutphen, United States Secretary of the Navy, Charles Edison, Rear Admiral Dunn, and Executive Vice President of Elco, Henry R. Sutphen. In Sutphen's hand is one of the programs handed out that night. (PT Boats Inc.)*

Figure 5-49. *Original Press book and entrance ticket. This is a five page press book with color photos, and inserts about the Elco Naval Division and their work in the building of the PT boats. Certainly one of the high points for not only the company, but the workers that made it all possible. These fine examples are from the authors personel collection. (Frank J. Andruss Sr.)*

Figure 5-50. *US Navy Officers and sponsors gather on the deck of this PT boat. The good people of Grand Haven, Michigan has purchased enough War Bonds and Stamps for one Elco boat. The plaque was presented to them and will be secured to the boats bulkhead. 4th from right in the first row is Lt. Cmdr. Frank W. Rowe, Elco's Assistant Supervisor of Shipbuilding. (PT Boats Inc.)*

Figure 5-51. *The two officers representing the Navy, secure the plaque from Grand Haven, Michigan inside the boats day room cabin. It will remain with the boat for as long as she is in the war zone. (PT Boats Inc.)*

Figure 5-52. *A ten cent dance is held at the American Legion in Bayonne. For ten cents you could get a dance with a pretty lady, and in this case the girls were workers at the plant. The money raised for this type of dance were placed into the war relief fund or to raise money for war bonds. Also a mock jail was set up and you had to get raised funds to get out of Jail. Lt. Cmdr. F. W. Rowe (PT Boats Inc.)*

Figure 5-53. *Some of the workers in the Machine Shop Assembly Department area, gather to show the Christmas tree they set up to celebrate the Holidays. There would be no work stoppage as production of the boats was paramount. Parties would be held in outside venues after work and workers would have a great time. (Frank J. Andruss Sr.)*

Figure 5-54. *These Bar tenders were on the committee called "Salvage For Victory" campaign. The campaign asked that all Americans do their part in supporting the new war effort by collecting, saving, conserving, and recycling materials that could be repurposed for military uses. In Bayonne and New York, mayors were asked to save the same kinds of materials from municipal dumps and incinerators. In Bayonne, the Department of Sanitation began picking up materials on an alternating schedule with regular garbage collection. In this photo they are being the nights special Bar Tenders to help raise money and awareness of the program. (PT Boats Inc.)*

Figure 5-55. *War bond dinner was held for Elco's Army/Navy E award presentation for excellence in productivity. It is not known who these gentlemen are but at least one of them, 3rd from left is wearing his Elco ID Badge. Notice the unopened bottle of wine on the table which has a label on it, showing one of the 80 foot Elco PT boats. This was also the night of the showing of the World Premier Featurette "Devil Boats" at the Dewitt Theater in Bayonne, New Jersey. (PT Boats Inc.)*

Figure 5-56. *As more men were leaving Elco for the service, women workers were being hired to replace the men. It was noted by management that outside activities for them were needed. Here women workers have formed bowling teams and competed the same as the men bowlers. (PT Boats Inc.)*

Figure 5-57. *One of the Women bowling teams was none other then the Mosquitoes. Other fun names for the ladies were the Sea Horses, the Eagles, and the Ravens. They held their games at local bowling allies, including the Dewitt Theater, which had a bowling ally in the basement. (PT Boats Inc.)*

Figure 5-58. *Great look at some of the skaters at the Boulevard Roller Rink. This was at the time located on the Boulevard at 62nd Street in Bayonne. Many Elco workers took the time to skate and have fun, but could also head here for Ball Room Dancing. During the war they actually had a campaign that you should skate to work to save gasoline for the war effort. (PT Boats Inc.)*

Figure 5-59. *Elco's Labor Management Team came up with what was called the Mop-Spruce-Cleanup contest. This was held plant wide, giving employees the chance to win prizes of War Bond Money and Certificates. These unknown Elco workers have been handed winning certificates from their respective department's for keeping their areas in tip top shape. Handing out the certificates is Mechanical Superintendent George Fulton. (Naval War College Archives)*

Figure 5-60. *On the lighter side of the war, Frederick Bulkeley (left) takes time out from his busy schedule to chat with his famous son, Lt. Cmdr. John D. Bulkeley. Frederick was the oldest employee at Elco as an inspector, and his son was famous for his exploits with a PT Squadron in the Philippines, that took General Douglas MacArthur from Corregidor and the escaping Japanese forces. Bulkely would be awarded the Medal of Honor. (Naval War College Archives)*

Figure 5-61. *Helen Jepson signs in at Elco at the Guard Station. Helen sang leading soprano roles with at Metropolitan Opera from 1935 to 1941. Some of her best known roles while at the Met include Desdemona (*Otello*) and Marguerite (*Faust*). The *Faust* recording is still in print, as is her recording of Porgy and Bess*.* She was the first soprano to record in that role, and the extant recording of her was supervised by Gerswin himself. She captured the hearts of opera lovers with her lyric soprano voice and radiant good looks. Helen was invited to sing for the workers as part of Labor Management's 4th of July entertainment. (Irwin Chase)*

Figure 5-62. *Managing Constructor Irwin Chase helps well known singer, Helen Jepson down the ramp, as her party gets set to board PT-544. Helen was treated to learn first hand how the boats operated. Later she would sing at the outdoor stage area. The boat would go on to serve with Squadron 37 in the Pacific. (Irwin Chase)*

Figure 5-63. *Helen Jepson takes to the outdoor stage area to address the workers and other guests. She will entertain them with her beautiful voice, singing some of her well known songs of the past. On the piano is Elco employee Earl McCurdy. Later Helen would work for the school district in Monmouth County, New Jersey as a speech therapist for children. When she retired she and Walter Dellera moved to Bradenton, Florida. She died at the age of 92 on September 16, 1997. (PT Boats Inc.)*

Figure 5-64. *Women from the Electrical Assembly Department take to the outdoor stage for some afternoon songs. On the piano was Earl McCurdy. This was another build up for purchasing War Bonds. Elco's motto was "A stamp is a bullet," a "Bond is a gun, Buy them both till the war is done". (Irwin Chase)*

Figure 5-65. *On August 10, 1942, Elco workers are all smiles after they took the company offer of War Bonds instead of vacations. In a company wide vote the majority of 2400 workers voted to giving up their vacation period to stay on the job and keep the production of PT boats at peak capacity. Soon after the vote, members of the Elco Employees Association held their 3rd annual outing at the Pavonia Yacht Club. Proceeds of the affair were used for the War Service Fund with all proceeds used to stock each PT boat leaving the plant, with cigarettes, razor blades, and chocolate. Mechanical Engineer George A. Fulton is on the far left smiling, while Elco employee 5th from left is Mike Banasiak. (Frank J. Andruss Sr.)*

Figure 5-66. *Managing Constructor Irwin Chase (left) shows off one of the creations made by Elco. This was a highly polished and detailed PT Boat hood ornament. It is not known exactly how many of these were produced, although one was used on the Elco company car, and one presented to Lt. Cmdr. John D. Bulkeley from Motor Torpedo Boat Squadron 3 fame, and recipient of the Medal of Honor. (PT Boats Inc.)*

Figure 5-67. *Beautifully detailed hood ornament has been attached to this 1941 Chrysler Crown Imperial vehicle. This was one of Elco's company car's and mostly used by Executive Vice President Henry R. Sutphen. (PT Boats Inc.)*

Chapter 6

End of an Era

When the war ended, the Elco Naval Division had over one thousand workers that had not only made PT BOATS but pleasure craft before that. Elco tried to keep them on the payroll but it was becoming very difficult. Elco struggled. In its woodworking facilities, Elco augmented yacht-building with other work, including manufacturing baseball bats tested by the Chicago Cubs during spring training one year, building bowling pins and wood lanes for bowling alleys, and crafting the wood work for the old "Woodies" a special edition Plymouth station wagon. By this time Elco was just a small branch of the much larger Electric Boat conglomeration, an empire later known as General Dynamics Corporation.

They built boats like the Eco thirty-five-foot Cruiser efficiently designed for easy riding, and claimed that it had a PT proven underbelly that cut hull resistance, that boosted speed per horse power. It had a fully equipped galley, lavatory, full size berths, and plenty of storage space below decks. All copper tanks and plumbing throughout the boat. These were offered in 1947 with five sizes from twenty-seven, thirty-five, forty, forty-seven, and sixty-two feet.

They built the beautiful top of the line 1947 sixty-two-foot Motor Yacht. A luxurious motor yacht with comfortable accommodations for seven guest's and two crew. This had a flying bridge, and was offered in both gasoline and diesel engines. The boat was beautifully finished with incredible attention to fittings, upholstery, and furnishings. African Mahogany throughout the boat, Eco spared no expense in making this one of the nicest yachts on the market.

Lacking proper support from what became General Dynamics's upper management, Elco was not given a chance to transition through the immediate post-war economy. By 1948, it was just a small branch of Electric Boat, a conglomerate handled by John Hopkins. Hopkins felt that Elco was not a profitable enterprise, and in December of 1949 the work force was let go, and the equipment, supplies and plant in Bayonne were sold.

After Elco left Bayonne, the structure was used by Englander Matress and Efka. On September 3, 2001, a fire caused severe damage at the former Elco Naval Division factory. The industrial structure was vacant at the time of the fire, and had not been active for several years. The fire would sweep thru the Elco complex destroying all but two of the nineteen buildings that would make up the site. It would also destroy three nearby homes and injure at least twenty-eight people. One of the largest casualties of the fire would be the Electro Dynamic Works, which at the time employed at least five hundred people in the manufacture of motors for submarines and other navy ships.

More then five hundred firemen battled this blaze for more then nineteen hours as flames as high as three hundred feet shot into the air. A thick layer of black smoke covered the city from bay to bay form Bergen Point to the Jersey City-Bayonne city line. Three New York City fire boats poured tons of water on the flames and were credited by local officials with saving the Englander Mattress Company from being destroyed, as well as the Elco Marina and the boats moored there. The fire was believed to have started around 2:00 p.m. on Saturday, April 20, 1963 in a small storage shed behind the Polytex building near Newark Bay.

It quickly engulfed the building touching off a series of explosions. Fanned by the winds coming across the Bay, the flames spread quickly to the Electro Dynamic buildings, the Englander warehouse area and three homes on Schulyer Court. The winds gusting to thirty-two miles per hour, tossed burning cinders across Avenue A and started fires on lawns and roofs along North Street and Avenue A.

The main fire would be fought by firemen from Bayonne, the Bayonne Naval Supply Center, Jersey City, Elizabeth, Weehawken, North Bergen and West New York One of the most dramatic moments of the fire was the rescue of seven firemen trapped by the flames at the end of the pier where they were manning pumps, drawing water from the bay.

They were saved by the Sea Scout Ship 7, skippered by Al Wiedow. This fire would mark the end of Elco as a complex, although building 21, the main assembly plant for the PT boats would be saved, as well as the dockside crane. Building 21 would later be completely destroyed and demolished some thirty-nine years later, when vandals set fire to it in 2002.

The engineer's report concluded that the main industrial building should be condemned and demolished. This recommendation applied only to the main structure. The Elco Marina clubhouse, the dockside crane, and other free-standing structures nearby were not harmed in the fire and were excluded from the recommended demolition.

The crane, recovered from the Elco Naval Division in Bayonne, lifted the PT boats from land and lowered them into Newark Bay. It serves as a reminder of the important role that Bayonne played in World War II, recalling the men and women employed by the Electric Launch Company (Elco), the servicemen who commanded and worked aboard the Patrol Torpedo (PT) boats that were built in Bayonne, and the innovation of the Electric Launch Company.

The Elco Crane is now in Richard A. Rutkowski Park at the Hackensack River Walk in Bayonne, on Newark Bay, west of 52nd–54th Streets (Block 12, Lot 3) In 2003 the crane was removed from its original location on Newark Bay to make way for the Boat Works housing development, which is located roughly two miles to the south of the park. The crane was dismantled and stored while this park was being completed.

Baker Residential had promised the people of Bayonne that it would restore the crane. On October 18, 2006 they made that promise come true as Bayonne city officials joined Baker residential and other invited guests for ceremonies. This special day was to dedicate the restored Elco Crane in its new site at Mayor Richard A. Rutkowski Park.

Once again the Crane stands tall, looking like she did when so many PT boats and pleasure craft were launched at Elco. Baker Residential also dedicated a plaque commemorating the Elco plant along the waterfront walkway, overlooking Newark Bay at the Boat-works site. Building 21, the main assembly building for the PT Boats was saved, as well as the attached offices of Building 20.

One thing is for certain Elco had a great business from its inception in 1892. It was run by Henry R. Sutphen for all but the first three of those years. Elco built pleasure craft in peacetime and large numbers of high-speed boats during both world wars. When the company started out, with a contract for fifty-five boats for the Chicago World's Fair, the hulls were subcontracted, but in 1902 they built their own yard in Bayonne, at the foot of North Street and extending northeast for six blocks.

Elco was the front runner in PT boat production, with six Army/Navy E awards being presented to the company for excellence in producing their boats for the war effort.

Elco strived to conduct themselves to the highest standard during the building of the PT boats. They worked tirelessly for a brighter future and better days after the war, in hopes of lasting peace. It was apparent that Elco its employees, and management filled an important niche, in the over-all production of the boats. The men who rode these wonderful boats did the Elco company proud, as Lt. John Bulkley won the Congressional Medal of Honor for his PT boat exploits during the opening days of the war. Lt. A. Murray Preston also won a Congressional Medal of Honor for his daring rescue of a downed flier two hundred yards from enemy gun positions in 1944. Twenty-two Navy Crosses were awarded to PT crewmen, along with hundreds of lesser citations.

Today, outside of the boatworks complex sits the Elco Marina. It is set up just about opposite where the main assembly building would have been during the war. It offers easy access to the Kill Van Kull River inlet, and is minutes away from the Bayonne Bridge. Baker Residential which built Boat Works complex, also remembered to honor those men and women that worked at Elco, with a wonderful plaque and memorial. It sits in a quiet area on the site with benches and retaining wall. Elco certainly made the people of Bayonne proud in what they accomplished during the war years. Hopefully the sacrifices that were made will never be forgotten. It is truly the end of an era.

Figure 6-1. *This worker has just opened the door to the oven, which has been heated to 140 degrees. It was left here for at least ten hours, making the wood soft and easier to form. This technique was the same used in the wood working process of the PT boats. (Frank J. Andruss Sr.)*

Figure 6-2. *Workers are building the jig for the frames that will be attached to make up the hull of one of the 30 foot pleasure craft. Looking above we see the mezzanine area on the second floor which was used to build cabin truck structures of the PT boats. (Frank J. Andruss Sr.)*

Figure 6-3. *Inside the main assembly building we see the center boat has been planked. It will soon will be hooked to the hoist to be turned right side up, an invention that was used during war time. To the lower left of the photo is a jig that will receive the frames that will make up the hull. Farthest forward to the right, we see an uncompleted PT boat. (Frank J. Andruss Sr.)*

Figure 6-4. *With the hull completed and hooked into over-head winches, it begins the turning process, so that workers can begin to work on the decking. Certainly not as large as the PT boats that once occupied this area, the assembly process to completion is much faster. (Andrew Shanahan)*

Figure 6-5. *With many of the pleasure craft much smaller in size, they could be towed out of Building 21 using the side garage door. Hooked into the tow motor and on her cradle on wheels the boat will make its way to the dockside crane. (Frank J. Andruss Sr.)*

Figure 6-6. *Now that the boat has left the assembly building, it has been hooked into the dockside crane. From here it will be gently lowered into the boat basin, and if other assembly is needed it will be towed to the wet boat basin. The front large sliding doors on rollers are kept closed to keep heat inside the assembly area. (Andrew J. Shanahan)*

Figure 6-7. *Now in the boat basin and floating on her own, the trial crew will begin to run tests to make sure all systems are running properly. She is one of Eco's 30 foot pleasure craft. (Andrew J. Shanahan)*

Figure 6-8. *The Eco 35' Cruiser was efficiently designed for easy riding and claimed that it had a PT proven underbelly that cut hull resistance, that boosted speed per horse power. It had a fully equipped galley, lavatory, full size berths, and plenty of storage space below decks. All copper tanks and plumbing throughout the boat. These were offered in 1947 with five sizes from 27, 35, 40, 47, and 62 feet. (Andrew J. Shanahan)*

Figure 6-9. *A look inside the 40' Cruiser Salon, showing the wonderful craftsmanship that Eco was known for. All trim was made from African Mahogany with a spacious helm made for comfort. The boat could sleep six people and had two lavatories, one with a shower. It had a roomy galley onboard and could hit a top speed of anywhere from 24 to 27 knots. (Andrew J. Shanahan)*

Figure 6-10. *Left Admiral Russel S. Berkey, USN Commamder of the New York Navy Yard, Mr. George W. Codrington, and Elco's Executive Vice President, Henry R. Sutpen aboard one of the Post War Elco 40' Express Cruisers. (PT Boats Inc.)*

Figure 6-11. *A beautiful port side look at Eco's top of the line 1947 62' Motor Yacht. A luxurious motor yacht with comfortable accommodations for seven guest's and two crew. This had a flying bridge, and was offered in both gasoline and diesel engines. The boat was beautifully finished with incredible attention to fittings, upholstery, and furnishings. African Mahogany throughout the boat, Eco spared no expense in making this one of the nicest yachts on the market. (Andrew J. Shanahan)*

Figure 6-12. *One can see the incredible workmanship that Eco put into the lounge area of it's 62' Motor Yacht. The furnishings are fit for a king and queen. Carpeted flooring with top of the line comfort built in with plenty of lighting and space for guests. (Andrew J. Shanahan)*

Figure 6-13. *Elco's 40' cruiser which offered top of the line furnishings with African Mahogany used throughout. Elco is using some very pretty models in bathing suits which made for a great selling tool. (Andrew J. Shanahan)*

Figure 6-14. *Port Elco located in Miami, Florida. This was used to showcase many of the Elco pleasure craft that were on display. Customers would be treated to a personal inspection of the boats both new and used. In 1923 Port Elco in New York was opened, and people could come off the street to inspect the boats. (PT Boats Inc.)*

Figure 6-15. *The offices of Port Elco in Miami, Florida. A rare look inside showing Elco workers in their every day job. The man on the left seems to be writing up a possible order for a boat, as the women in the chair listens. (PT Boats Inc.)*

Figure 6-16. *Port Elco in Miami had a wonderful lounge area for customers, complete with a working stone fire place. On the easel to the right is a poster of Elco's 1940 57 foot Yacht. (PT Boats Inc.)*

Figure 6-17. *The beautiful complex of the Electro Dynamic Company with the first floor being occupied by the Elco Works of the Electric Boat Company. The fire in 1963 would destroy these entire buildings. (Frank J. Andruss Sr.)*

Figure 6-18. *The massive heat and wind swept fire destroyed almost everything in its path. Twisted steel and burned out buildings were all that was left. All but two of the 19 buildings that made up Elco were lost. Suffering a great deal was the Electro Dynamic Company, that at the time of the fire employed 500 people that manufactured motors for submarines and other navy ships. (Andrew J. Shanahan)*

Figure 6-19. *Electro Dynamic workers left: Morris Hagen and Sal Marino sift through the rubble to try and save anything that could be used for future use. They are trying to collect castings. (Andrew J. Shanahan)*

Figure 6-20. *Fire has destroyed what was once the Elco vault located on the second floor of this building. To the left you can see Building 20, the office area which was not harmed by the fire, nor was the attached Building 21, the main assembly area for the boats. (Andrew J. Shanahan)*

Figure 6-21. *A crane with wrecking ball has begun to demolish what was left of the structures for Elco and Electro Dynamics. It would take several months to clear out this area. Fire-Fighters still had to contend with hot spots within the wreckage. (Andrew J. Shanahan)*

Figure 6-22. *In 1998 Bayonne resident and former Elco worker Frank Karahuta, invited me to visit the site. This was one of the photo's I took when I first arrived after walking out to the docks. Seeing this building was one of the highlights of my life. It is easy to see the neglect of not only the building, but the concrete roadway in front which has deteriorated and parts of it have fallen into the water. (Frank J. Andruss Sr.)*

Figure 6-23. *A nice look at the Twin Boom crane on the concrete roadway. This crane was built at Bergen Point Iron Works in Bayonne, and was a reminder as to the boats that were gently lowered into the boat basin. The crane has a sling set up for lowering boats into the water. To the left of the crane are some storage buildings, on the dock, and to the right the white building was the Elco Club house. (Frank J. Andruss Sr.)*

Figure 6-24. *The inside of the main assembly building was shown to me by Elco Marina Manager John Tencza. Although not in terrible shape, the roof had begun to rot and rain and the elements would puddle on the floor. Electric to the building had been shut off. You can see some of the original green funnel lights that were present during the building of the PT boats. (Frank J. Andruss Sr.)*

Figure 6-25. *Another look at the assembly building looking towards the back. You can clearly see the water on the floor, and debris that has fallen on the floor, which has been placed into piles. (Frank J. Andruss Sr.)*

Figure 6-26. *The dockside crane would continue to operate while the Marina was on site. The crane control house was original to the crane during the time the PT boats and pleasure craft were being launched. The crane ran by electricity and heavy cables. It should be noted that the crane was never painted yellow during the war years, but was gray as was the control house. (Frank J. Andruss Sr.)*

Figure 6-27. *This is the long view from the dock to the exit, which would lead to Newark Bay. These are the original break water pylons that you see just outside of the docks that are in the water. This view is looking out from the main assembly building. (Frank J. Andruss Sr.)*

Figure 6-28. *A second spectacular second fire in September 2001, would light up the sky over Bayonne's Bergen Point section around 8:00pm on a Monday. This fire would seal the fate of the old building that was a prominent figure during the years of making the PT boats. From this inside photo you can see the incredible damage that was done, as one wall has collapsed and the portion of this roof caved in. (Andrew J. Shanahan)*

Figure 6-29. *Another look inside the assembly building after the fire. Fire officials would label the blaze suspicious. There were no utilities in the building, nothing that could be determined as a source of ignition. (Andrew J. Shanahan)*

Figure 6-30. *Ladder truck (2) from Bayonne, fights the spectacular blaze that could be seen for many miles. Ultimately some 50 fire-fighters and almost every fire company in the city helped pour water into the burning structure. No fire-fighters would enter the building, as it was deemed too unsafe. (Andrew J. Shanahan)*

Figure 6-31. *Fire-fighters on an extension ladder pour water on many of the hot spots the day after the fire. When they arrived on the scene the night before, they would find heavy flame and smoke coming through the building. (Andrew J. Shanahan)*

Figure 6-32. *Demolition of the site is in full swing, and the only structure's left standing are the Elco crane, and the Elco club house. In 2003, under the recommendation of the Historical Society of Bayonne, the crane would be dismantled and stored for future use. (Andrew J. Shanahan)*

Figure 6-33. *A brick from Building 21 was sent to the author by Bayonne resident Andy Shanahan, as a reminder of a once proud company. (Frank J. Andruss Sr.)*

Figure 6-34. *After the success of removing debris from the fire, the land was purchased and surveyed. On the old Elco site and beyond, some 32 acres of land would be used to create many waterfront town houses in a gated community, called Boatworks. These wonderful homes would offer spectacular views and give people a chance to live close to New York. (Frank J. Andruss Sr.)*

Figure 6-35. *This photo shows the Townhouses, which sit very close to the main entrance of Building 21, the main assembly plant for PT boats and later pleasure craft. To the right is the Elco Marina. (Frank J. Andruss Sr.)*

Figure 6-36. *A beautiful memorial that was erected in 2005, sits inside the boat-works community. It pay's tribute to those men and women that worked at the Elco site. It also pays tribute to the officers and enlisted men that served aboard the PT boats during WWII. This area is also lit at night and has park benches for those wishing to sit and reflect in this honored area. (Frank J. Andruss Sr.)*

Figure 6-37. *Just outside of the boatworks complex sits the Elco Marina. It is set up just about opposite where the main assembly building would have been during the war. It offers easy access to the Kill Van Kull River inlet, and is minutes away from the Bayonne Bridge. (Frank J. Andruss Sr.)*

Figure 6-38. *The white topped retaining wall around the Boatworks Town Houses pretty much shows what was once the Elco Naval Division and Elco Yacht Works facilities. You can clearly see the original break water pylons to the right of the Elco Marina Docks. (Frank J. Andruss Sr.)*

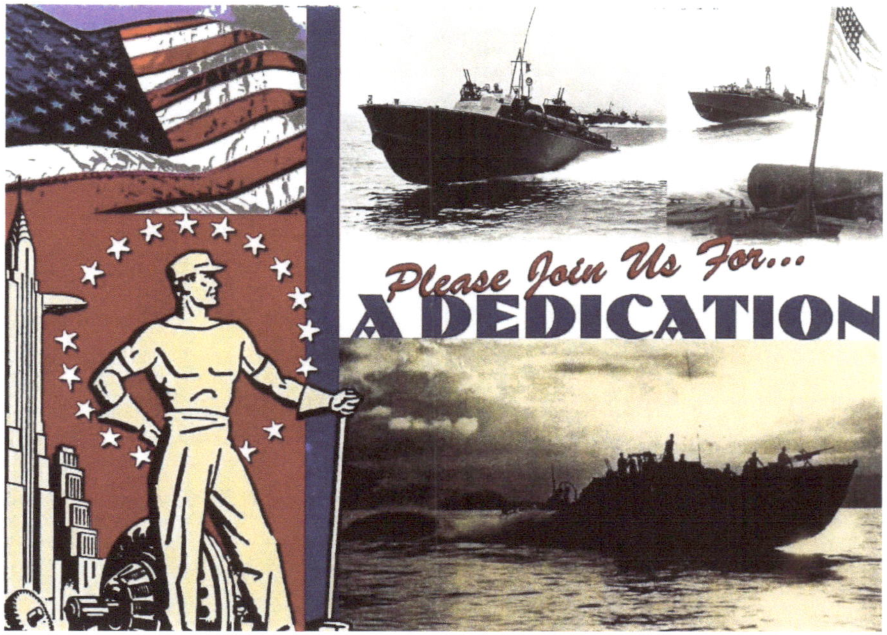

Figure 6-39. *On October 4, 2006 a dedication for the restored Elco Crane took place. Guests invited by Mayor Joseph V. Doria Jr., the City Council of Bayonne, and Baker Residential would met at the site, which was chosen for the this historical crane, and her final resting place. The site chosen was inside the Mayor Richard A. Rutkowski Park, in Bayonne, New Jersey. (Andrew J. Shanahan)*

Figure 6-40. *Guests at the Crane Dedication Ceremony are Korean War Veterans Left: Andrew J. Shanahan, Larry Berulli, and Barry Dugan. The fourth man is WWII Disabled American Veteran, Henry Sanchez. (Andrew J. Shanahan)*

Figure 6-41. *A wonderful look at the crane which looks out over the water, not far from where she began her career with Elco. Today she proudly stands waiting for another PT boat to gently drop into the water, only this time on a peaceful mission from God. She is here to remind visitors of her past and the work she did rushing the little wooden boats to the war zone. (Frank J. Andruss Sr.)*

Figure 6-42. *From the Elco crane dedication ceremony, this paper weight was given to the Author by Bayonne, New Jersey resident Andrew J. Shanahan. It is a wonderful reminder of the dedication by so many that helped not only save the crane but restore it for future generations to enjoy. (Frank J. Andruss Sr.)*

Acknowledgements

First and foremost I have to say thank you to my wife Stacia, who once again has put up with my antics for a fifth time. I love her more than anything, and know that my endless stories about the PT boats have taught her that I will never stop loving this passion of mine. Nice to have her with me for the ride.

As always I must give a big thank you to those men and women who worked three shifts each week, to produce these wonderful Elco PT boats. They came together at a crucial time, as rushing these boats to the war front was a monumental task. Anytime one does pictorial books, one has to make sure enough photos of the subject matter are out there, so first and foremost special thank you to PT Boats Inc. for their many photos.

Special thanks goes out to my friend Andy Small, who took the time to scan many of these photos. Without his assistance this book goes nowhere. I am so thankful that Andy always goes the extra mile to help me when I need it. I am extremely thankful for his assistance in the research and development chapter of the book, and the travels he made to scan much needed photos. Special thank you to the Naval War College Archives that provided photos from their collection. Thanks to my friends Jerry Gilmartin, Charlie Jones, Al Ross, and Ted Walther, who I can always count on to help me with photos, history, and keep me flying straight. Special thanks to PT Boats Inc, and their photo archives that appear in this book.

Special thanks to my friends Alex Johnson, Bruce Chip Marshall, and John Molnar, who are watching me from Heaven, making sure I am continuing to help educate those that know little of the PT boats of World War II. You guys are greatly missed. Thanks to my friend Andy Shanahan, who has lived in the Bayonne area for many years, and was an employee of Electro Dynamic Company. His friendship, photos and history of the area mean a great deal.

Special thank you to John Cunningham for the many photos provided by his grandfather Irwin Chase, the Managing Constructor of the Elco Naval Division. John was gracious in opening up his photographic treasures so we could copy many of these photographs for this book. Heather Rowe for her donation of photos from her Father, Frank W. Rowe, the Assistant Supervisor for the Elco Naval Division, Jim Waller for photos from his dad, Charles Waller who was an executive assistant to the Works Manager, at the Elco Naval Division, Al Ross for his wonderful color illustrations, and Morris Rosenfeld, photographic Illustrator for the Elco Naval Division.

Thanks to Gene Kirkland for not only providing some photos for this book, but his knowledge as well. Thanks to Bert Morgan, Time Life Magazine, Kingston Winget, Donald Rosencrantz, and the Naval History and Heritage Command, for their photos. Thank you to James Gray who helped me fix a few photos. Thanks to those who worked at Elco in the photo Lab, producing these wonderful photos, many years ago.

Special thanks to my Mom, who at ninety-one years of age, still stands behind me as I ramble on about the history of the PT boats, and to my sisters Debbie and Lori who always are willing to help me when it comes to my passion of Elco and the PT boats. Love you all. If I have forgotten anyone, please forgive me.

Finally to our four Grandchildren, Dominick, Frankie, Charlie, and Oscar. It remains my hope that you will one day take an active interest in the History of the PT boats and that maybe on some cloudy or rainy day, you will take one of my books, in a comfortable chair and read it.

www.ingramcontent.com/pod-product-compliance
Lightning Source LLC
LaVergne TN
LVHW071515220926
848486LV00009BA/375